SCALAR DIFFRACTION FROM A CIRCULAR APERTURE

SCALAR DIFFRACTION FROM A CIRCULAR APERTURE

CHARLES J. DALY
Rochester Institute of Technology

NAVALGUND A. H. K. RAO
Rochester Institute of Technology

Kluwer Academic Publishers
Boston/Dordrecht/London

Distributors for North, Central and South America:
Kluwer Academic Publishers
101 Philip Drive
Assinippi Park
Norwell, Massachusetts 02061 USA
Telephone (781) 871-6600
Fax (781) 681-9045
E-Mail <kluwer@wkap.com>

Distributors for all other countries:
Kluwer Academic Publishers Group
Distribution Centre
Post Office Box 322
3300 AH Dordrecht, THE NETHERLANDS
Telephone 31 78 6392 392
Fax 31 78 6546 474
E-Mail <services@wkap.nl>

Electronic Services <http://www.wkap.nl>

Library of Congress Cataloging-in-Publication Data

Crisis and change in the Japanese financial system / edited by Takeo Hoshi,
Hugh Patrick.
     p.cm. --(Innovations in financial markets and institutions)
  Includes bibliographical references.
  ISBN 0-7923-7783-4 (acid-free paper)
   1.Finance--Japan.2.Financial institutions--Japan. I.Hoshi, Takeo. II.Patrick,
Hugh T. III. Series.

HG187.J3 C75 2000
332'.0952--dc21

                                       00-020395

Printed on acid-free paper.

Printed in the United States of America

Contents

List of Figures

Preface

The problem of scalar diffraction from a circular aperture is ubiquitous, fundamental, and enduring. It is ubiquitous because it naturally arises in a variety of disciplines, such as optics (lenses), acoustics (speakers), electromagnetics (dish antennas) and, of course, ultrasonics (piston transducers). The problem is fundamental because of its simple description coupled with its not-so-simple set of solutions—solutions which have preoccupied scientific minds since 1885 at least. The problem endures, despite centuries of research, because generation after generation rediscovers it and adds to the existing literature some novel insight or new result. This monograph promises a few new results and many novel insights. Although the text emphasizes ultrasonic diffraction, the results and insights developed are general and may be applied to any problem involving scalar diffraction from a circular aperture.

Thus, mathematicians, scientists, and engineers working in optics, astronomy, acoustics, antenna design, biomedical engineering, and nondestructive testing should find the development, results, and insights presented in this monograph both interesting and useful. The material is based on a synthesis of mathematics, physical optics, linear systems theory, and scalar diffraction theory. As a result, the text is suitable for graduate courses in applied mathematics, science, and engineering or for self-study and reference.

This monograph has three goals. The first and foremost is to derive a closed-form spatially-averaged two-way diffraction correction for a focused ultrasonic piston transducer operating in pulsed mode. The first goal is attained by (i) establishing that, in the case of one-way diffraction with a point receiver, the arccos and Lommel diffraction formulations form an approximate Fourier transform pair and (ii) exploiting this approximate Fourier equivalence in rigorous and original derivations of spatially-averaged diffraction corrections for both one-way and two-way diffraction. In addition to the first goal, the monograph has two larger goals. The first is to develop a new perspective on spatially-averaged diffraction correction for piston transducers. The second is to advance the scientific community's understanding of scalar diffraction from a circular aperture. When viewed in its entirety, the monograph attains all three goals by presenting rigorous derivations and original analyses that unify and extend existing theory with new insights and generalized results.

This work is dedicated
to my family,
Rose of Sharon,
Mister Charlie, and
especially my beloved
wife, Yon Hui, and to
my parents, who taught
me the value of an
education. *CJD*

To Lakshmi, Rohit and
Priti whose love and
support made this
possible, and to my
parents. *NHR*

Acknowledgments

This research had three goals. The first was to derive diffraction corrections for focused piston transducers. The second was to develop a new perspective on the theory of spatially averaged diffraction corrections. The last was to advance the scientific community's understanding of scalar diffraction from a circular aperture. We believe we have accomplished these goals. Many people contributed to this accomplishment, and we wish to acknowledge their contributions.

Dr. Joseph Hornak provided valuable criticism and suggested pedagogical improvements to the layout of this document. Dr. Vince Samar asked challenging questions. Indeed, his questions changed the direction of our research for the better. Dr. Roger Easton inspired us with his unique perspective on life and linear systems theory. Discussions with him streamlined at least one of our derivations. Dr. María Helguera, our close colleague, provided stimulating discussion and warm friendship. Dr. Kevin Parker and other members of the Rochester Center for Biological Ultrasound weighed in with cogent comments and constructive criticism. Finally, special thanks go to Alex Greene, our publisher.

Chapter 1

INTRODUCTION

This work has three goals. The first and foremost is to derive a closed-form spatially averaged two-way diffraction correction for a focused piston transducer operating in pulsed mode. It is attained by (i) establishing that, in the case of one-way diffraction with a point receiver, the arccos and Lommel diffraction formulations form an approximate Fourier transform pair and (ii) exploiting this newly established Fourier equivalence in rigorous and original derivations of spatially averaged diffraction corrections for both one-way and two-way diffraction. The second goal is to develop a theoretically and historically unified perspective of spatially averaged diffraction corrections for piston transducers. The third goal is to advance the scientific community's understanding of scalar diffraction from a circular aperture. When viewed in its entirety, this research attains all three goals by presenting rigorous derivations and original analyses that unify and extend existing theory with novel insights and generalized results.

In addition, the theory is applied to gauge its practicality. Specifically, two-way diffraction corrections are applied to ultrasonic data obtained from laboratory experiments. The diffraction corrections are time-varying filters implemented with a short-time Fourier technique known as the *weighted overlap-add* method [22]. Raw and diffraction-corrected RF data are quantitatively compared via spectral centroids, and B-mode images are reconstructed from diffraction-corrected data.

The problem of scalar diffraction from a circular aperture is ubiquitous, fundamental, and enduring. It is ubiquitous because it naturally arises in a variety of disciplines, such as optics (lenses), acoustics (speakers), electromagnetics (dish antennas) and, of course, ultrasonics (piston transducers). The problem is fundamental because of its simple description coupled with its not-so-simple set of solutions—solutions which have preoccupied scientific minds since 1885 at least. The problem endures, despite centuries of research, because generation after generation rediscovers it and adds to the existing literature some novel insight or new result. This monograph promises a few new results and many novel insights. Although the text emphasizes ultrasonic diffraction, the results and insights developed are general and may be applied to any problem involving scalar diffraction from a circular aperture.

What is scalar diffraction from a circular aperture, and why bother looking at a problem that has been investigated off and on for over a century? What do terms such as *spatially averaged*, *two-way*, and *pulsed-mode* mean and imply? Why are diffraction corrections important? How does diffraction from a circular aperture arise in the context of ultrasound?

The purpose of this chapter is to answer these and other questions and, in so doing, introduce concepts and definitions used in subsequent chapters. The emphasis at this stage is on pedagogy rather than precision. Also included in this chapter is a section which delineates the scope of the research; the same section also states underlying assumptions. Another section previews the remaining chapters, and a final section addresses potential objections to the proposed theory.

1.   ULTRASONIC REFLECTION IMAGING

In a biomedical context, an ultrasonic transducer generates and couples a sound wave to the anatomy under examination (Fig. 1.1). The sound wave interacts with tissue and is scattered as it propagates. Some of the scattered energy reflects back to the transducer. The acoustic energy of the reflected wave is received and converted into a voltage that varies as a function of time. The voltage at the output of the receiver shown in Fig. 1.1 is known as an *A-line*.

An A-line is a radio-frequency (RF) voltage whose spectrum is centered about the stated operating frequency of the transducer. The RF A-line can be analyzed at RF or, more commonly, it can be brought to baseband via a process known as *envelope detection* and subsequently displayed as an intensity line. This is known as *A-mode* imaging [4]. Envelope-detected or baseband data is considered low-frequency data, while RF data is considered high-frequency data. More details on baseband and RF data can be found in the literature on communications theory [33].

The transducer can be moved laterally, and a 2-D array of RF A-lines built up. Each RF A-line in the 2-D array can be processed via envelope detection and subsequently displayed as a gray-scale intensity

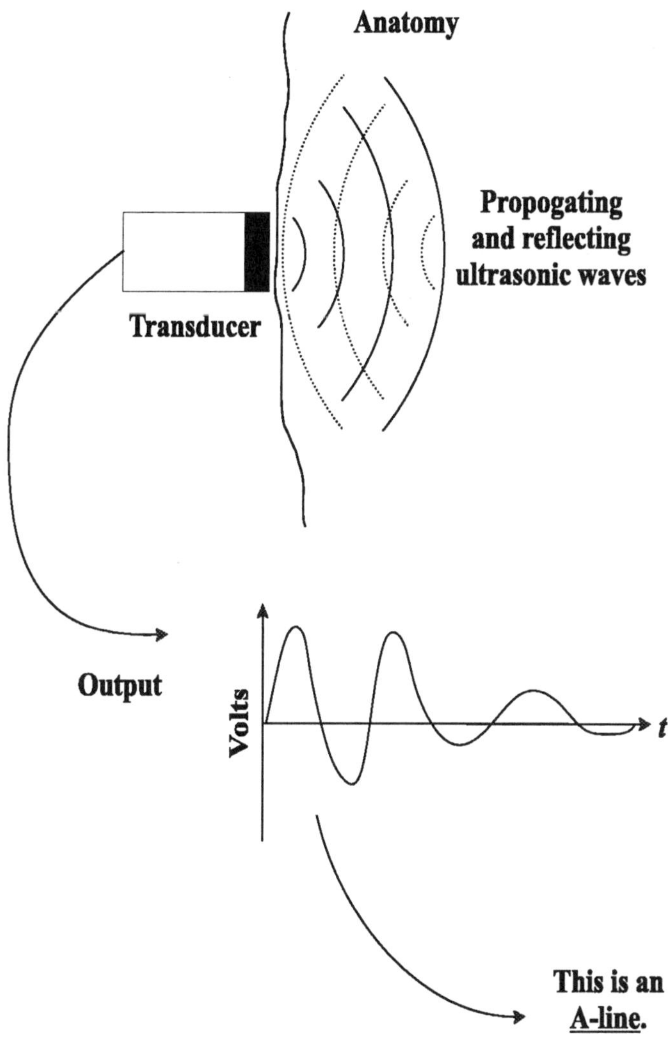

Figure 1.1.   Ultrasonic reflection imaging.

image. This is known as *B-mode* imaging. Gray-scale obstetric images
are familiar examples of B-mode imaging. Both A-mode and B-mode
reflection imaging are similar to radar. The transducer sends out ultrasonic energy and detects what energy is reflected back. Like an antenna
in monostatic radar, the transducer in ultrasonic reflection imaging acts
as both transmitter and receiver.

In terms of efficiency and cost, ultrasonic imaging is much faster and cheaper than other imaging modalities such as X-ray and magnetic resonance imaging (MRI). But in terms of resolution and contrast, ultrasonic imaging is vastly inferior to these same imaging modalities. One cause for the inferior quality of ultrasonic imaging is diffraction, the topic of the next section.

2.   DIFFRACTION FROM A CIRCULAR APERTURE

The phenomenon known as *diffraction* is fundamental to the study of wave propagation through an aperture of any shape. However, only circular apertures are considered in this research. Diffraction from a circular aperture can be described with a simple, yet familiar, example from optics. Consider shining a flashlight on a wall in a dark room. The illumination on the wall depends on the distance between the flashlight and the wall—the larger the distance between the flashlight and the wall, the dimmer the illumination but the greater the illuminated area; the smaller the distance, the brighter the illumination but the smaller the illuminated area. More succinctly, brightness is inversely proportional in some fashion to distance, while illuminated area is proportional in some fashion to distance. Fig. 1.2 illustrates this familiar example and depicts diffraction as beam or energy spread. Diffraction affects acoustic and ultrasonic energy in a similar fashion. A loudspeaker and piston transducer are quintessential examples of diffraction from a circular aperture in acoustics and ultrasonics, respectively.

The important point is that diffraction causes waves to spread in a spatially varying fashion. In reflection imaging, distance or depth z is related to time t via $z = ct/2$ where c is the speed of sound. Thus, the effects of ultrasonic diffraction change with time, and correcting for the effects of ultrasonic diffraction will require time-varying filtering of some kind. This simplistic characterization of diffraction as spatially varying beam spread becomes inadequate when other factors, such as frequency, aperture size, and focusing, are considered.

For monochromatic or single-frequency excitation of a *given frequency*, the distance from the aperture at which noticeable beam spread begins is proportional in some fashion to the aperture size. Monochromatic excitation is often called *continuous wave excitation* in the ultrasound community. Similarly, for a *given aperture size*, the distance from the aperture at which noticeable beam spread begins is proportional in some fashion to the excitation frequency. Focusing adds more complexity. Fig. 1.3 illustrates how changes in aperture size, frequency, and focusing affect diffraction from a piston transducer.

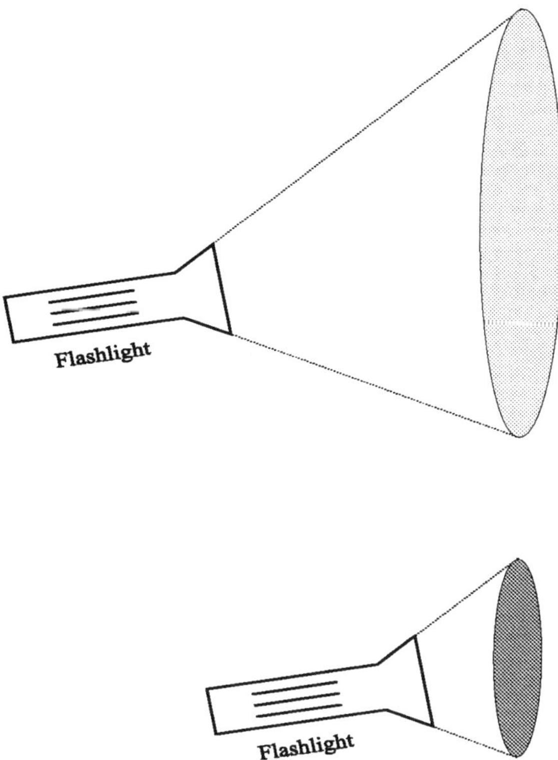

Figure 1.2. Simple example of diffraction.

Fig. 1.3 implies that the beam is approximately collimated out to a certain distance and then begins to diverge. The notion of a clear-cut demarcation between collimated and diverging regions is a useful oversimplification and applies to monochromatic excitation only. Furthermore, the collimated and diverging regions are loosely associated with the *near field* and *far field*, respectively [34].

A piston transducer operating in pulsed mode is said to be excited polychromatically or impulsively; the term *polychromatic* implies multi-frequency or pulsed excitation; Describing the diffraction effects becomes more complicated when the transducer is excited polychromatically. In the polychromatic case, there are collimated and diverging regions associated with each frequency contained in the excitation. The importance of these observations on the near field, far field, and excitation type will be clarified in subsequent chapters.

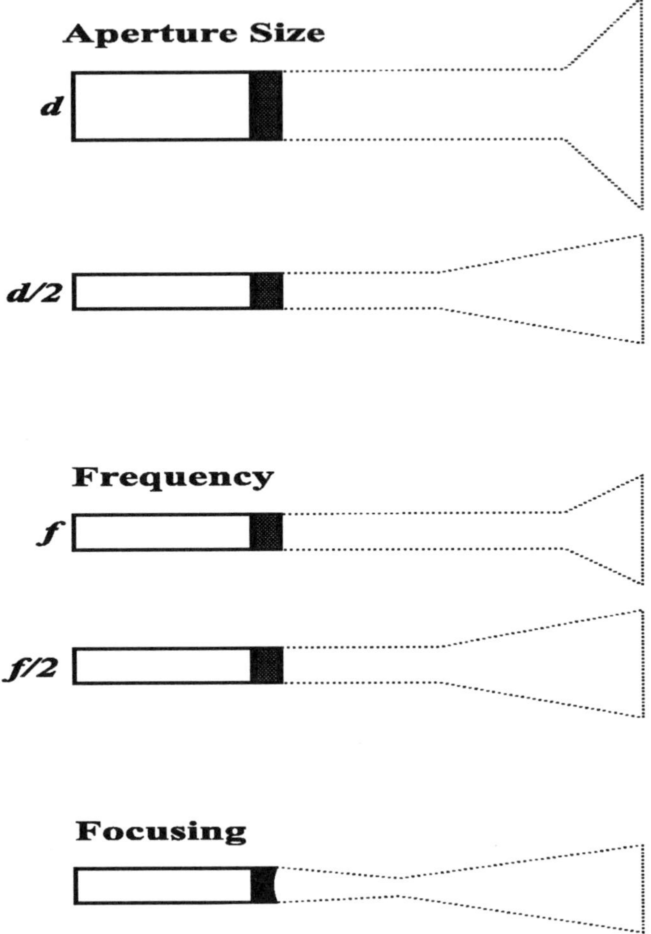

Figure 1.3.   Factors affecting diffraction.

3.   THE ARCCOS & LOMMEL DIFFRACTION FORMULATIONS

The discussion of diffraction can be put on a more mathematical footing by considering Fig. 1.4 and introducing the scalar function $H_1(\rho, z, \omega)$ which characterizes the disturbance sensed by a point receiver at a radial distance ρ from the axis due to some monochromatic excitation of the aperture. It can be shown [36] that $H_1(\rho, z, \omega)$ satisfies the time-independent Helmholtz wave equation:

$$(\nabla^2 + k^2)H_1(\rho, z, \omega) = 0. \tag{1.1}$$

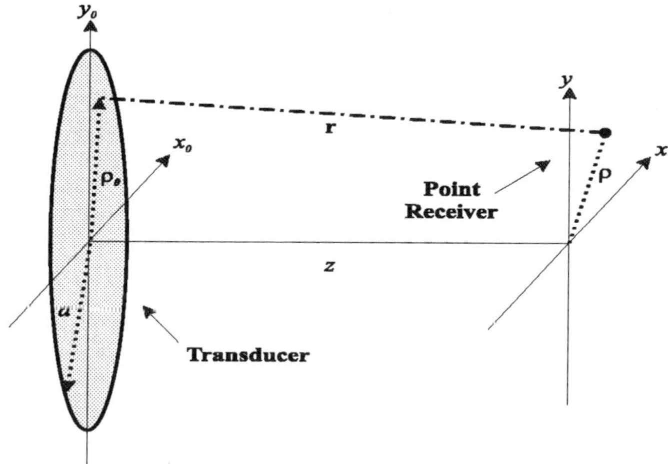

Figure 1.4.   Diffraction from a circular aperture.

Lommel [38] investigated this problem in 1885, and his solution for $H_1(\rho, z, \omega)$ is the Lommel diffraction formulation. In 1961, Oberhettinger [66] characterized the disturbance at the point ρ due to pulsed or impulsive excitation of the aperture, and his solution is the arccos diffraction formulation. As an important aside, the ultrasound community has considered the piston transducer problem in terms of diffraction from a circular aperture since at least the 1940's [45].

For now, attention is drawn to the fact that the arccos and Lommel diffraction formulations are similar in that they both characterize the effects of *one-way* diffraction from a circular aperture at a *point* (an infinitesimally small area); thus, the two formulations seem amenable to some type of unification. Much more will be said about the arccos and Lommel formulations and their connection in Chapter 3. The distinction between one-way and two-way diffraction and its importance is explained in the next section.

4. ONE-WAY AND TWO-WAY DIFFRACTION

From a physical standpoint, one-way diffraction implies energy travel in one direction—away from the source. Two-way diffraction implies energy travel in two directions—first away from the source and then back to source after interaction with matter. The physical distinction is illustrated in Fig. 1.5 for a point receiver and point scatterer. This

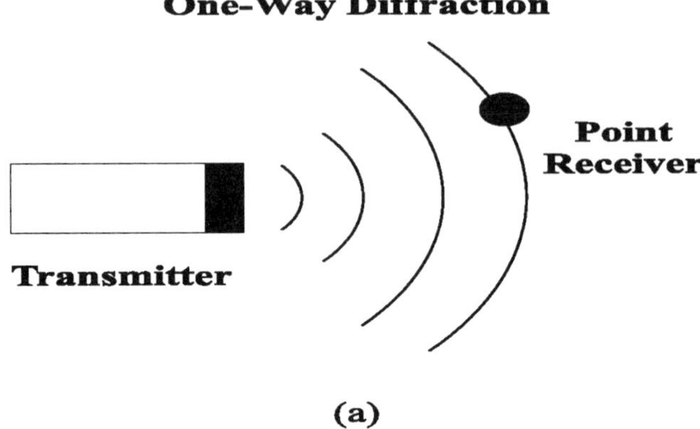

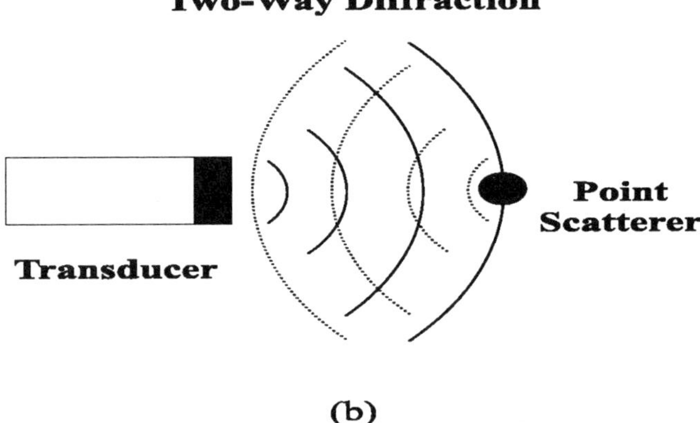

Figure 1.5.   (a) One-way diffraction with point receiver; (b) two-way diffraction with point scatterer.

distinction is particularly important in ultrasonic reflection imaging because it is subject to mathematical interpretation.

Specifically, some authors state that equations derived for one-way diffraction can be used to calculate two-way diffraction from an infinite plate simply by doubling the distance between the source and infinite plate in the one-way equations [16, 76, 91]. This claim is based on a *mirror-image* interpretation of two-way diffraction from an infinite plate. It is important to note that this mirror-image interpretation holds only in the case of two-way diffraction involving a infinite plate.

On the other hand, some authors state that two-way diffraction is properly described by the square of $H_1(\rho, z, \omega)$ [30, 44, 93]. This claim is based on an interpretation of two-way diffraction as an autoconvolution. Both mirror-image and autoconvolution interpretations have merit and mathematical appeal. Because of this, closed-form spatially averaged equations applicable to both interpretations will be derived in later chapters.

A note on terminology is required at this point. The physical distinction between one-way and two-way diffraction coupled with the mirror-image and autoconvolution interpretations of two-way diffraction leads to a rich but potentially confusing taxonomy. The confusion arises because the term *two-way diffraction* is imprecise; it may imply either the mirror-image or autoconvolution interpretation of two-way diffraction. Here, the term *two-way diffraction* is associated with the autoconvolution interpretation. The terms *mirror-image diffraction* and *autoconvolution diffraction*, although awkward, will be used when the mathematical interpretation of two-way diffraction requires specification. More will be said about mirror-image and autoconvolution diffraction in Chapter 7.

5.   SPATIAL AVERAGING

The discussion up to this point has focused on fictitious point receivers and point scatterers. The receiving area of a real transducer, however, is finite; thus real transducers are often referred to as *finite receivers*. Fig. 1.6 depicts more realistic scenarios for one-way and two-way ultrasonic diffraction. In the one-way case, an ultrasonic transducer emits energy and another transducer coaxially located with the transmitter some distance z away acts as a receiver.

The usual goal of one-way ultrasonic probing is to extract information about the medium between the transmitter and receiver. This information is encoded on the RF A-line. Recall that an A-line is the output voltage of the receiver and is illustrated in Fig. 1.1. The A-line is assumed to be a function of either the total pressure or spatially averaged pressure impinging on the face of the receiver [5, 14, 41, 96]. Williams makes the case for total pressure, while Harris makes the case for spatially averaged pressure. In either case, total pressure is found by spatial integrating the incoming pressure field over the face of the receiving aperture; note that the receiving aperture is referred to as a *measurement circle* by Williams [95]. Spatially averaging consists of simply dividing the total pressure by the receiver area. Thus, spatial averaging and spatial integration differ by a constant multiplicative factor.

Spatial averaging of one-way ultrasonic diffraction has been investigated extensively, and numerous authors have used one-way results to

One-Way Diffraction

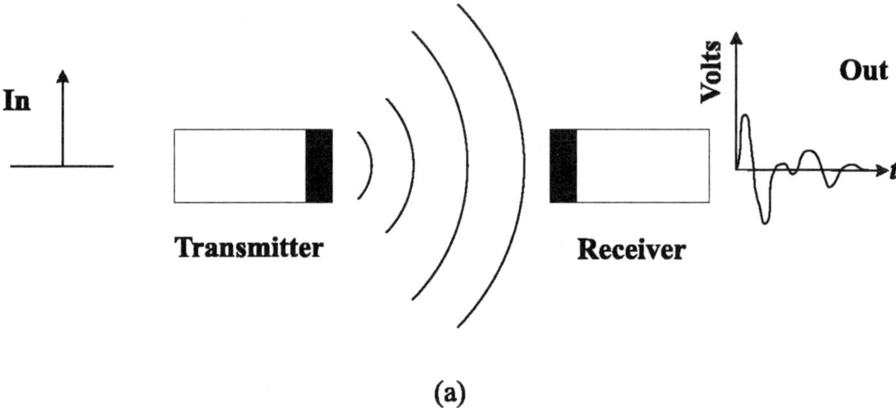

(a)

Two-Way Diffraction

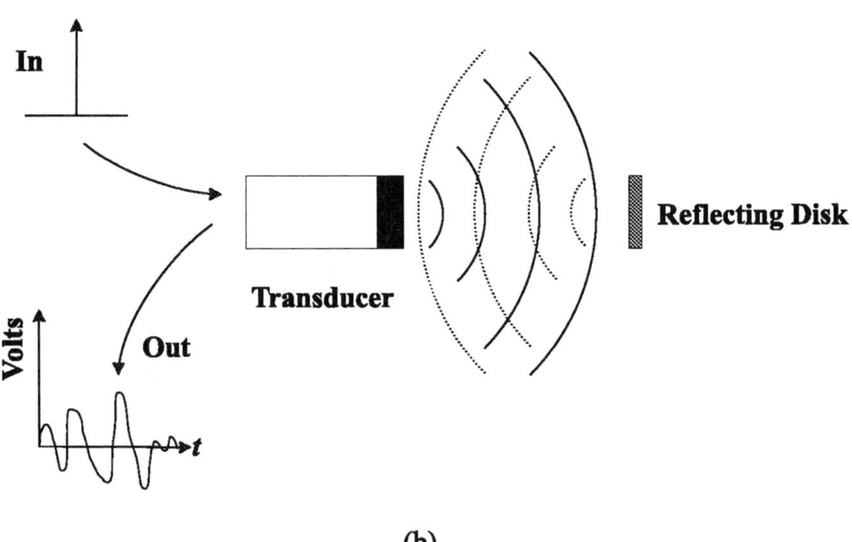

(b)

Figure 1.6.   (a) One-way diffraction with finite receiver; (b) two-way diffraction with finite reflecting disk.

explain two-way diffraction from a infinite flat plate by invoking the mirror-image interpretation of reflection imaging. In the autoconvolution interpretation of reflection imaging, diffraction is modeled for both transmission and reflection. In this case, total pressure is found by spatial integrating the incoming (reflected) pressure field over the receiving aperture; the reflected pressure, in turn, is found by spatially integrating the outgoing (transmitted) pressure field over some suitably chosen reflecting plane [13, 30]. Averaging consists of dividing by the area of the receiving aperture.

Spatially averaged autoconvolution diffraction from a flat plate has not been investigated as extensively as spatially averaged one-way diffraction, Reference [13] being a notable exception. Chapter 5 considers autoconvolution diffraction involving a flat plate with new tactics; there it is empirically shown that closed-form results derived by Wolf in 1951 for optical diffraction can be used to estimate a spatially averaged form of autoconvolution diffraction from a flat plate. This new but relatively straightforward empirical proof is based on the Cauchy-Schwarz inequality for integrals [71, pp. 177-178] and an *ad hoc* assumption concerning its interpretation. Despite the *ad hoc* and empirical development, useful results are obtained.

6. THE NEED FOR DIFFRACTION CORRECTION

Terms such as *spatially averaged*, *two-way*, and *pulsed-mode* have been defined and their importance explained in the context of ultrasound. The phenomenon of diffraction from a circular aperture and its applicability to ultrasonic reflection imaging have been considered. The question of why a century-old problem is being re-investigated remains. The inescapably long answer can be found in an area of ultrasonic research known as *tissue characterization* [29, 58]. The answer also explains the need for diffraction correction.

Tissue characterization is the Holy Grail of ultrasound because radiologists are already highly skilled at *characterizing* abnormal tissue with other imaging modalities such as and computed tomography (CT) and MRI [73]. These other imaging modalities are, however, more expensive and time-consuming than ultrasound. Furthermore, the harsh economics of managed care are forcing radiologists to cut costs at every turn and, as a result, to order fewer CT and MRI work-ups. Consequently, the medical community at large has compelling economic need to develop tissue characterization methods that are quick, reliable, and affordable. The Holy Grail of ultrasonic tissue characterization, when and if perfected, can help fulfill that need.

As was just mentioned, radiologists are already very good at characterizing abnormal tissue when they have recourse to more expensive and time-consuming imaging modalities. They are not yet as adept at characterizing abnormal tissue when limited, as managed care would prefer, to diagnosis via ultrasound. One of the reasons radiologists are not adept at classifying abnormal tissue is their almost exclusive reliance on B-mode *images* when limited to ultrasonic diagnoses. Recall, B-mode images are constructed from low-frequency envelope-detected data.

Thus, a radiologist examining a B-mode image does not "see" the whole ultrasonic picture; he or she is making a radiological diagnosis based primarily on low-frequency envelope-detected data. Any potentially important diagnostic information encoded on the high-frequency RF data is lost as a result of envelope detection. Researchers in ultrasonic tissue characterization are trying to develop new tools and methods that help radiologists see this potentially important diagnostic information encoded on the high-frequency RF data.

Some of these new methods of tissue characterization involve spectral analysis of RF data [32, 84]. Tissue characterization based on spectral analysis combines diagnostic information from low-frequency envelope-detected data—B-mode images— with additional information available from the high-frequency data—RF A-lines. Advanced digital signal processing offers many methods of obtaining this additional information from the high-frequency data.

The results produced by these spectral methods are, however, biased by the transmit and receive impulse responses of the transducer and two-way diffraction effects. The bias can be removed by spectral normalization or diffraction correction [37, 91]. As Lizzi and co-workers note, spectral normalization "is performed by dividing measured tissue spectra by the calibration spectrum obtained from the front surface of a glass plate located in a water tank [59]" at some distance z from the transducer. Note well that the calibration spectrum must be obtained experimentally for each distance z of interest.

It is theoretically possible to obtain the spectral normalization required for any depth z with only one calibration spectrum measurement. The procedure is as follows. Obtain the calibration spectrum for some depth z. Compute the diffraction effects, also known as acoustic or radiation coupling [14, 16], for the same depth and divide them out from the calibration spectrum. What is left is the combined transmit/receive response of the transducer. This response is time invariant; hence, it can be combined with a computation of two-way diffraction effects to arrive at the complete spectral normalization for any depth z of interest. This discussion helps explain the need for diffraction correction and

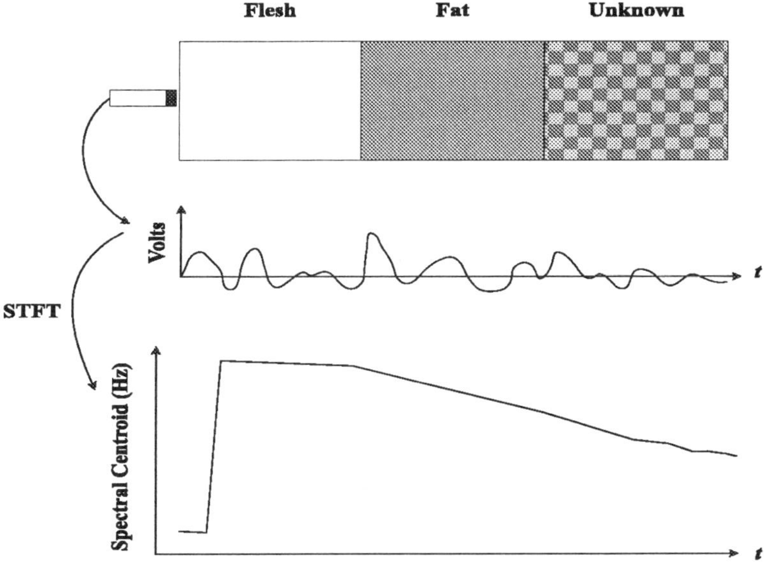

Figure 1.7.   Tissue characterization.

reminds us of the first and most immediate goal of this monograph: to derive a closed-form spatially averaged two-way diffraction correction for a focused piston transducer operating in pulsed mode.

An example may help make things a bit more concrete. Fig. 1.7 illustrates a notional example. For simplicity, only one A-line is considered.

The figure shows three layers of tissue and the RF A-line obtained by probing the layers with a piston transducer operating in pulsed-mode. A measure of the average or dominant frequency contained in the RF A-line—its spectral centroid— can be obtained via short-time Fourier processing [32]. A notional spectral centroid is shown in the figure.

The spectral centroid varies in some quantitative fashion that correlates, at least theoretically, with the type of tissue being probed. This theoretical correlation, however, is biased by two-way diffraction effects. Specifically, the spectral centroid is based on high-frequency RF data, but diffraction, as will be shown, attenuates high-frequency information. Thus, the spectral centroid will be biased by high-frequency attenuation unless, as explained earlier, the diffraction effects are corrected.

We have ignored the issue of spatial averaging in our discussion of the need for diffraction correction. Physically meaningful correction of two-way diffraction effects for finite receivers requires spatial averaging.

Derivation of a spatially averaged autoconvolution diffraction correction using the arccos diffraction formulation [86] was attempted but proved too difficult. This difficulty prompted our investigation of the Lommel diffraction formulation, a solution to the century-old problem of monochromatic diffraction from a circular aperture. As will be shown, deriving and implementing a spatially averaged autoconvolution diffraction correction turns out to be fairly straightforward once the Fourier equivalence of the arccos and Lommel diffraction formulations is established.

A note on terminology is required at this point. Note that the term *diffraction correction* can be interpreted as the total removal of unwanted diffraction effects. We do not make this interpretation. The time-varying filters derived here simply compensate for depth-dependent, high-frequency attenuation caused by diffraction, and the term *correction* is used loosely. Furthermore, the distinction between diffraction *effects* and diffraction *correction* is simply a matter of inversion, and the two terms are often used interchangeably in the literature on diffraction correction. The same semantic liberty is taken in this monograph.

In summary, diffraction attenuates high-frequency information and biases tissue characterization; hence, a correction is required to remove the bias caused by diffraction. We derive equations describing diffraction effects as a set of time/depth-varying filters. These diffraction filters are simply inverted to obtain diffraction corrections.

7.   MATHEMATICAL DEFINITIONS

The development of spatially averaged diffraction corrections for piston transducers is made possible, in part, by a group of related mathematical functions which will surface repeatedly in later chapters. These functions are defined here. Following Wolf, u and v are real variables, n and m are non-negative integers, and $J_n(u)$ is a Bessel function of the first kind of order n. $U_n(u,v)$ and $V_n(u,v)$ are Lommel functions of two variables and are defined by summations of Bessel functions [97]:

$$U_n(u,v) = \sum_{s=0}^{\infty} (-1)^s \left(\frac{u}{v}\right)^{n+2s} J_{n+2s}(v), \qquad (1.2)$$

$$V_n(u,v) = \sum_{s=0}^{\infty} (-1)^s \left(\frac{v}{u}\right)^{n+2s} J_{n+2s}(v). \qquad (1.3)$$

Because they are infinite summations, the Lommel functions can be computed only approximately, and these approximations can be programmed either recursively [38, 97] or directly in a do-loop. Do-loops were used in this work.

$U_n(u, v)$ converges slowly when $u/v > 1$, so the following formulae from Gray and Mathews [38, p. 185, Eq. 20] will prove useful:

$$U_{2n+1}(u, v) + V_{-2n+1}(u, v) = (-1)^n \sin\left(\frac{1}{2}\left[u + \frac{v^2}{u}\right]\right), \qquad (1.4a)$$

$$-U_{2n}(u, v) + V_{-2n}(u, v) = (-1)^n \cos\left(\frac{1}{2}\left[u + \frac{v^2}{u}\right]\right). \qquad (1.4b)$$

Special-case formulae for $u/v - 1$ can be found in Gray and Mathews.

A group of related functions will also be encountered:

$$Z_n(u, v) = (-1)^s U_{n+2s}(u, v), \qquad (1.5)$$

$$= \sum_{s=0}^{\infty} (-1)^s (s + 1) \left(\frac{u}{v}\right)^{n+2s} J_{n+2s}(v);$$

$$W_n(u, v) = (-1)^s V_{n+2s}(u, v), \qquad (1.6)$$

$$= \sum_{s=0}^{\infty} (-1)^s (s + 1) \left(\frac{v}{u}\right)^{n+2s} J_{n+2s}(v);$$

$$X_n(u, v) = \frac{1}{2}\left[\frac{v^2}{u}U_{n+1}(u, v) + uU_{n-1}(u, v)\right], \qquad (1.7)$$

$$= \sum_{s=0}^{\infty} (-1)^s (n + 2s) \left(\frac{u}{v}\right)^{n+2s} J_{n+2s}(v) \quad \text{and}$$

$$Y_n(u, v) = \frac{1}{2}\left[\frac{v^2}{u}V_{n-1}(u, v) + uV_{n+1}(u, v)\right], \qquad (1.8)$$

$$= \sum_{s=0}^{\infty} (-1)^s (n + 2s) \left(\frac{v}{u}\right)^{n+2s} J_{n+2s}(v).$$

$Z_n(u, v)$ was encountered in the course of this work, and it is discussed in an appendix to Chapter 4. More on details on $W_n(u, v)$, $X_n(u, v)$, and $Y_n(u, v)$ and their origins and applications can be found in [97, 98]. Additionally, the polynomials

$$P_{n,2m}(v) = \sum_{s=0}^{2m} (-1)^s J_{n+2}(v) J_{n+2m-s}(v) \qquad (1.9)$$

$$Q_{2m} = P_{0,2m}(v) + P_{1,2m}(v) \qquad (1.10)$$

will be encountered. It is important to note that all the equations introduced in this section can be easily coded by anyone even only mildly proficient in modern programming languages.

Finally, an important historical disclaimer is required. With the exception of perhaps $Z_n(u, v)$, all functions introduced in this section are due to Lommel, Hopkins, or Wolf. See Gray and Mathews [38], Watson [90], and Wolf [97, 98] for more details. Indeed, the closed-form equations derived in subsequent chapters were made possible, in large part, by results presented in Wolf's 1951 paper on diffraction; the importance of Wolf's results to this work cannot be overstated.

8.   SCOPE AND ASSUMPTIONS

This section describes the scope of the proposed work and explains some choices and assumptions that have been made. More details will be discussed as the need arises. This monograph focuses on A-mode and B-mode biomedical ultrasonic reflection imaging using linear pulse-echo techniques. M-mode, pulsed-Doppler, color-flow, tomographic, and more exotic ultrasonic imaging modalities are not discussed. In short, the scope of this research is limited to linear, non-Doppler, pulse-echo reflection imaging. However, non-linear ultrasound is discussed briefly in Chapter 7.

Ultrasound experiments were conducted with piston transducers which were assumed to be infinitely baffled. The term *infinitely baffled* means that the piezoelectric membrane of the piston transducer is restrained by an infinitely rigid wall [50, 66]. More will be said about equipment in Chapter 6. Array transducers are briefly discussed in Chapter 7.

Scalar diffraction theory is assumed throughout. Specifically, Rayleigh-Sommerfeld diffraction theory forms the mathematical foundation for the results that will be derived [36]. Furthermore, an infinitely baffled ideal transducer with infinite bandwidth or Dirac response is assumed in the theoretical development [14]. The first Born approximation, which implies no multiple scattering, is also assumed [50]. In short, we assume ideal linear ultrasonic propagation in a uniform, isotropic, homogeneous and weakly scattering medium which supports only compressional waves [76]. These assumptions concerning ultrasonic diffraction and linear propagation are ubiquitous in the literature.

As was already mentioned numerous times, diffraction can be modeled as a time-varying filter. Practical time-varying filtering requires a joint time-frequency representation that is invertible *and* realizable with undue computational burden. Most joint time-frequency representations in the current literature can be conveniently classified as linear, bilinear/quadratic, or non-linear, and only a few meet the practical requirements just described [42]. Indeed, non-linear representations are generally not invertible, and they were not considered. Bilinear representations were considered but abandoned because they are not easily invertible.

Ultimately, the short-time Fourier transform was used for time-varying filtering because it is linear, invertible, well-documented [65], and easily interpreted. The short-time Fourier transform is viewed here as a tool, not a subject of study in and of itself. Hence, no tutorial material on the subject is included. However, references to tutorial material and seminal works on the short-time Fourier transform and short-time Fourier techniques are included.

9.   PREVIEW

Recall the three theoretical goals of this work. The immediate goal is to derive a closed-form spatially averaged two-way diffraction correction for a focused piston transducer operating in pulsed-mode. The second goal is to connect, with coherent and generalized results, previously disjoint theory on spatially averaged diffraction corrections for piston transducers. The third goal is to advance the scientific community's understanding of scalar diffraction from a circular aperture. This monograph is organized to achieve these goals, and we preview it here.

Chapter 2 reviews relevant literature and gives more background material. Chapter 3 is probably the most important chapter in the document for three reasons. First, it introduces and compares the arccos and Lommel diffraction formulations [23]. Second, Chapter 3 establishes and verifies the Fourier equivalence of the arccos and Lommel diffraction formulations as an approximate Fourier transform pair for both focused and unfocused piston transducers. In short, the two diffraction formulations are connected in a new way thus providing fresh insight into diffraction from a circular aperture. Third, the new results and insights developed in Chapter 3 form the mathematical foundation for a new treatment of spatially averaged diffraction corrections for an infinitely baffled ultrasonic piston transducer operating in pulsed mode.

This new treatment or perspective is based on the hypothesis that spatial averaging in the time domain is the same as spatial averaging in the frequency domain for any diffraction problem that can modeled with the Rayleigh-Sommerfeld diffraction integral [36]. In essence, our development is a frequency-domain formalism for spatially averaged diffraction correction. As such, it is a frequency-domain alternative to the well-established time-domain or impulse-response formalism for spatially averaged diffraction correction [14, 76, 86].

One-way spatially averaged diffraction effects are derived in Chapter 4. The insight gained in Chapter 3 leads to a new derivation of closed-form spatially averaged time-domain expressions based on the arccos diffraction formulation [14, 24]. The derivation relates ultrasonic and optical diffraction in a unique way and thus provides a different perspective on diffraction from a circular aperture. In addition to this novel derivation, two new closed-form frequency-domain expressions are derived by spatially averaging the Lommel diffraction formulation [23, 25]. An appendix to Chapter 4 contains two lemmas helpful in deriving the spatially averaged Lommel formulation.

Numerical results obtained from the time-and frequency-domains expressions are analyzed. The Fourier equivalence of the arccos and Lommel diffraction formulations for point receivers predicts that results obtained from the spatially averaged arccos-based results should be approximately equal to results obtained by inverse Fourier transforming the Lommel-based expressions; in short, the two formulations should be equivalent in a Fourier sense. This prediction is verified, and the success of the theory in the case of one-way diffraction with point and finite receivers is cause for optimism that theory will also hold for spatially averaged autoconvolution diffraction.

Spatial averaging of two-way diffraction *involving a flat plate* is considered in Chapter 5. Specifically, a set of equations derived for optical diffraction is applied to the autoconvolution interpretation of two-way ultrasonic diffraction from a flat plate. The equations, which are based on the Lommel-Wolf treatment of Fresnel diffraction from a circular aperture, are completely general in terms of area and focusing. Results obtained by numerical integration of the arccos diffraction formulation are compared to results obtained from the closed-form Lommel-based equations. As in the one-way case, the Fourier equivalence of the arccos and Lommel diffraction formulations predicts that the two sets of results should be equivalent in a Fourier sense. The prediction is verified.

Thus, Chapters 3–5 show that the approximate Fourier equivalence of the arccos and Lommel diffraction formulations leads to a wholly new perspective on the theory of spatially averaged diffraction corrections for an infinitely baffled ultrasonic piston transducer operating in pulsed mode. The theory is unified in the sense that the Fourier equivalence of the two formulations applies to both one-way and two-way ultrasonic diffraction with piston receivers and reflecting disks of any size. The theory applies to unfocused and focused transducers and is used to derive closed-form results for one-way and autoconvolution diffraction, both focused and unfocused.

An aspect of the proposed formalism, autoconvolution diffraction, was experimentally investigated, and the results obtained are presented in Chapter 6. Specifically, autoconvolution diffraction corrections were implemented with time-varying filters, and diffraction-corrected B-mode images will were reconstructed using a short-time Fourier analysis and synthesis algorithm known as the weighted overlap-add (WOLA) method. The raw and diffraction-corrected images are compared only qualitatively. Differences between raw and corrected RF data are analyzed quantitatively via spectral centroids described in the previous section.

Three points concerning the experiments require discussion. First, it must be emphasized that the experiments were not designed to validate the theory in any authoritative fashion. Rather, they were designed to gauge the feasibility of the proposed autoconvolution diffraction correction. Second, the diffraction-corrected images are an important contribution of this work. Although the differences between the raw and diffraction-corrected images are subtle, they reveal that diffraction correction appears to have a more pronounced effect on RF data than on envelope-detected data. Finally and most importantly, the experiments are not to be considered, in any way, clinical validation of the proposed diffraction corrections.

Chapter 7 develops more results and provides deeper insight into the problem of scalar diffraction from circular aperture. Specifically, linear and non-linear ultrasound are discussed briefly, a focused one-way result is derived, coherent and incoherent averaging are compared, and the mirror-image and autoconvolution interpretations of diffraction are discussed. The final chapter draws conclusions and makes recommendations for further study.

10.   CRITICISM AND COUNTER

Those familiar with this area of research may have two obvious and immediate objections to the theory to be developed. The first objection is that the theory applies only to piston transducers which are the exception rather than the rule in ultrasound. The second objection is that the proposed theory will suffer from Gibb's phenomenon. These objections must be addressed at the outset.

The first objection is addressed with five short points. First, it must be re-emphasized that the results derived herein have applicability in other disciplines such as optics (lasers), acoustics (speakers), and electromagnetics (dish antennas). Second, the hypothesis that spatial averaging is the same in the time and frequency domains is general because it holds for any transducer geometry. Third, closed-form spatially averaged solutions for piston transducers are worthwhile in and of themselves and in terms of the theoretical insight gained by deriving them. Fourth, results describing diffraction from piston transducers are qualitatively and quantitatively useful in understanding diffraction from more general transducers [3, 18]. Finally, piston transducers remain an economically viable product in both biomedical ultrasound and non-destructive testing [69]. Thus, our monograph should be of theoretical and practical interest.

Addressing the second objection requires more explanation. The proposed frequency-domain formalism is based on the Lommel diffraction formulation and, as such, serves as an alternative to the well-established time-domain or impulse-response formalism based on the arccos diffraction formulation [14, 86]. Thus, we are primarily interested in frequency-domain results, particularly for spatially averaged diffraction effects.

To establish the formalism, we assume impulse excitation of an infinitely baffled piston transducer that has a Dirac response. To validate the formalism, we inverse Fourier transform frequency-domain results and compare them with results computed using the time-domain formalism. This comparison invites a third objection because results computed using the impulse-response formalism have finite support in the time domain.

In other words, our proposed frequency-domain formalism will suffer from Gibb's phenomenon because an infinite bandwidth can never be adequately sampled [101]. Thus, the proposed formalism will never exactly agree with the time-domain formalism. The criticism is valid for theoretically ideal transducers; however, we will have compellingly demonstrated the validity of our proposed frequency-domain formalism if it agrees reasonably well with the time-domain formalism. We will show that it does. Furthermore, real transducers are band-limited. Thus, the proposed formalism will be of practical value if it can be computed easily and accurately across bandwidths representative of real transducers. We will show that it can.

Chapter 2

LITERATURE REVIEW

This chapter reviews relevant literature on ultrasonic reflection imaging, diffraction from a circular aperture, the arccos and Lommel diffraction formulations, ultrasonic diffraction, short-time Fourier techniques, and time-varying filters. Literature on other topics, such as linear models of ultrasonic reflection imaging and the mirror-image and autoconvolution interpretations of ultrasound, will be discussed in subsequent chapters.

1. ULTRASONIC REFLECTION IMAGING

Ultrasonic reflection imaging is described in many sources; we discuss only four of these sources. Shung's treatment of ultrasound includes sections on acoustic propagation and diagnostic methods [82]. Macovski devotes two chapters to biomedical ultrasound in his text on medical imaging, one each on ultrasonic imaging theory and array transducers [60]. Macovski's development is based on linear system theory and will appeal to electrical engineers, while Shung's is more relevant to mechanical engineers. Thijssen's development of ultrasonic reflection imaging focuses on texture analysis and image processing but also provides a good introduction to basic ultrasound [88]. Finally, the chapter on ultrasound by Bamber and Tristam [4] in Reference [92] covers a broad range of ultrasound topics by emphasizing results and applications rather than mathematical development.

Taken together, these four references provide a fairly thorough introduction to ultrasonic reflection imaging and simultaneously point the way to more advanced references. For completeness, we note that the references just discussed all treat ultrasound from a primarily *linear* perspective. Works by Christopher and Hamilton & Blackstock, for example, can be used as a starting point for research into the burgeoning field of *non-linear* ultrasound [18, 39].

2.   DIFFRACTION FROM A CIRCULAR APERTURE

During his investigation of optical diffraction from a circular aperture in the late 1800's, E. Lommel developed a mathematical description of diffraction which we call the *Lommel diffraction formulation*; the description includes two celebrated functions which now bear Lommel's name [97]. After its publication in 1885, Lommel's work caught the attention of the mathematical community. Much of Lommel's original work was redone by Gray and Mathews in the early 1920's [38], and Watson [90] investigated Lommel functions from a mathematical standpoint at about the same time as Gray and Mathews. In short, the mathematical community refined the Lommel diffraction formulation.

The ultrasound community appears to have become interested in the Lommel diffraction formulation in the late 1940's when Huntington, *et al.* mentioned it in a paper on ultrasound delay lines and spatially averaged diffraction effects [45]. Eight years later, Seki, *et al.* published a paper on Lommel functions and diffraction effects in the case of monochromatic excitation of a piston transducer which they modeled as a circular aperture [81]. In the early 1950's, the ultrasound community began developing pulse-echo techniques [70], and the monochromatic Lommel diffraction formulation was applied in this new area of ultrasound for a number of years. What was needed, however, was a closed-form time-domain solution.

This solution was first derived by Oberhettinger in 1961, and we call it the *arccos diffraction formulation* [66]. His derivation is an analytical one based on Bessel theory and the Laplace transform. A decade later, Stepanishen re-derived the arccos diffraction formulation using arguments from physics and geometry [86]. Both derivations model the piston transducer as a circular aperture. In 1976, Penttinen and Luukkala derived a focused version of the arccos diffraction formulation [72]. Five years later, Arditi, *et al.* extended Penttinen and Luukkala's work by describing transient fields of concave annular arrays [2]. In that same year, Harris published a comprehensive review of the development of diffraction theory for pulsed piston transducers; his review includes the work of Oberhettinger, Stepanishen, and others [40]. Beginning in the 1980's, research shifted from the arccos diffraction formulation to spatially averaged diffraction corrections.

3. SPATIALLY AVERAGED DIFFRACTION CORRECTIONS

The ultrasound community has been researching spatially averaged diffraction corrections for piston transducers for over 50 years. The early work of Huntington, *et al.* has already been mentioned [45], and research in this area was pursued well into the 1950's. Williams derived one of the first closed-form diffraction corrections for an unfocused piston transducer in 1951 [95]. His derivation applies to a receiver or measurement circle equal in area to the transducer and centered in the transducer beam. In 1958, Bass [5] derived a closed-form result which was slightly more compact than the result derived by Williams.

There was also interest in spatially averaged diffraction effects outside of the ultrasound community. In the same year that Williams published his closed-form diffraction correction, Wolf extended Lommel's treatment of Fresnel diffraction and derived expressions "for the fraction of the total illumination present within certain regions in receiving planes near focus of spherical waves issuing from a circular aperture ... [97]" In short, Wolf used the Lommel diffraction formulation to find the spatially integrated intensity impinging on a disk coaxially located some distance z from the aperture. Wolf placed no restrictions on the area of the illuminated disk, and his many results may be used to spatially average both one-way and two-way ultrasonic diffraction.

In the 1960's, the ultrasound community focused its attention on the arccos diffraction formulation and its theoretical and experimental validity as a velocity-potential impulse response involving a convolution integral [40]. Work on spatially averaged diffraction corrections resumed in the 1970's when Williams published a paper [96] extending Bass's 1958 one-way diffraction correction to the case of a receiver having an area different from that of the transmitter. With computers becoming more accessible in the 1970's, researchers began to explore the validity and utility of the closed-form diffraction corrections derived in the previous two decades. For example, Khimunin [49] and Benson and Kiyohara [7] computed one-way diffraction corrections numerically and presented their results in tabular form. In fact, Benson and Kiyohara based their algorithms on Seki's 1956 paper [81].

In 1974, Rogers and Van Buren [79] simplified Bass's 1958 result by spatially integrating the Lommel diffraction formulation. Four years later, Rhyne [76] derived a closed-form one-way diffraction correction by spatially integrating the arccos diffraction formulation; he presented closed-form results in both the time and frequency domains. The results derived in References [79] and [76] are limited to transmitters and receivers having equal areas. In 1981, Harris discussed spatially averaged diffraction corrections for the case of arbitrary velocity distributions [41]. Two years later, Kuc and Regula computed spatially integrated diffraction effects via numerical integration and investigated their impact on spectral estimates for ultrasonic tissue characterization [55]. It is critical to note that all the authors mentioned to this point invoked the mirror-image interpretation of two-way diffraction and applied one-way results to two-way diffraction.

In 1983 and 1984, Fink, *et al.* published papers on diffraction effects in pulse-echo measurements [31, 32]. Their work was new in that they addressed the autoconvolution interpretation of two-way diffraction and introduced spectral centroids as a measure of diffraction effects for focused and unfocused piston transducers. In 1988, Cassereau, *et al.* [14] generalized Rhyne's one-way results [76] to transmitters and receivers having unequal areas. In the 1990's, Fink and Cardoso [13] derived a closed-form autoconvolution diffraction correction via spatial integration of a joint time-frequency representation. Madsen and Zagzebski have also computed diffraction corrections in their various papers on backscatter coefficients [15, 62]. In 1994, Chen *et al.* derived a spatially integrated mirror-image diffraction correction based on the Lommel diffraction formulation for the case of a focused piston transducer [16].

Ultimately, spatially averaged diffraction corrections are inextricably intertwined with the topic of ultrasonic reflection and scattering, an exceedingly difficult topic which in general is not amenable to analytic solutions. More will be said about this in later chapters. For now, we direct the reader to Dickinson [28], Cho, *et al.* [17], and the references these authors cite for detailed discussion of ultrasonic reflection and scattering.

4.   SHORT-TIME FOURIER TECHNIQUES

A *stationary* signal is one with time-independent spectral content; a *non-stationary* signal is one with time-varying spectral content [77]. Traditional Fourier techniques characterize stationary signals in either the time domain or frequency domain; joint time-frequency information is not readily available in either domain. In general, traditional techniques are not well suited to determining the time-varying spectra of non-stationary signals. The theory of joint time-frequency representations was developed to extend the applicability of traditional Fourier techniques to non-stationary signals. Hlawatsch and Boudreaux-Bartels wrote an excellent tutorial on joint time-frequency representations, and, as the authors point out, researchers in signal processing have concocted a plethora of joint time-frequency representations to analyze non-stationary signals [42].

The joint time-frequency representation that is most relevant to this research is the short-time Fourier transform (STFT), which Gabor proposed in 1946 [42]. Since then, the STFT has been used extensively in speech processing, and researchers in speech processing have developed many useful techniques based on the STFT. The chapter on short-time Fourier analysis in Reference [74] is an older reference but remains one of the best introductions to the subject. More recent work on the short-time Fourier transform and related techniques can be found in Nawab and Quatieri's contribution to Lim's 1988 text on advanced signal processing [65, 57] and Oppenheim and Schafer's text on signal processing [68].

Filtering of stationary signals is relatively straightforward because their spectral content is independent of time. On the other hand, filtering non-stationary signals is difficult because their spectral content varies with time. This type of filtering requires time-varying filters which may be implemented using short-time Fourier techniques. Thus, short-time Fourier techniques find wide application in time-varying filtering.

Since time-varying filters have been researched at least since the 1950's, a great deal of discussion of the theory can be found in the signal processing literature. Bello characterized randomly time-variant linear channels using time-varying filters, and his work provides a good introduction to the topic [6]. During the 1960's and 1970's, researchers in the speech community made great strides in time-variant filtering specifically because of the STFT and its linearity and invertibilty. See Lim [57], and Oppenheim and Schafer [68], Rabiner and Schafer [74], and the numerous references contained therein for more details. The time-varying filtering done in this work is based on the weighted overlap-add method of short-time Fourier analysis/synthesis developed by Crochiere in the late 1970's while he was working with Bell Laboratories [22].

5. SHORT-TIME FOURIER TECHNIQUES IN ULTRASOUND

Because diffraction can be modeled as a time-varying filter, short-time Fourier techniques have been applied in ultrasound. Salomonsson and Bjökman used a parametric time-varying network based on the STFT to separate attenuation and texture due to tissue [80], while Claesson and Salomonsson used the STFT to compensate for frequency- and depth-dependent attenuation in ultrasound signals [20]. More recently, Daponte, *et al.* [26] compared the STFT with other joint time-frequency representations in measuring the thickness of thin multilayer structures.

Outside the biomedical community, Malik applied different joint time-frequency representations, including the STFT, to the problem of ultrasonic non-destructive testing [63].

Much of the ultrasound literature discussing the STFT is in the area of Doppler signal processing. A recent example is Reference [89] in which the authors compare Doppler signal analysis techniques in the measurement of velocity, turbulence, and vortices; among the methods investigated was the STFT. Other representative work on short-time Fourier techniques in ultrasound was done by Altes and Faust [1] and Fink, *et al.* [31, 32]. Altes and Faust used short-time Fourier analysis to provide a unified framework for ultrasonic diagnosis, and the work of Fink, *et al.* has already been discussed.

6.   CHAPTER SUMMARY

This chapter reviewed relevant literature on ultrasonic reflection imaging, diffraction from a circular aperture, the arccos and Lommel diffraction formulations, ultrasonic diffraction, short-time Fourier techniques, and time-varying filters. Rarely is a literature review ever fully complete. Hence, we have made a good faith effort to provide the reader with an adequate survey and update-to-date review of scalar diffraction from a circular aperture as it pertains to ultrasound.

Chapter 3

TWO DIFFRACTION FORMULATIONS

This chapter establishes and verifies the Fourier equivalence of the arccos and Lommel diffraction formulations as an approximate Fourier transform pair. This relationship is important because it serves as the mathematical foundation for a proposed frequency-domain formalism of spatially averaged diffraction corrections for ultrasonic piston transducers. Although the development is cast in terms of ultrasonic propagation, the results are applicable to any physical problem involving scalar diffraction from a circular aperture. Some of the material in this chapter first appeared in References [23] and [25], and it is reprinted here with permission.

First, derivations of the arccos and Lommel diffraction formulations are outlined, and the two formulations are compared. The reader is referred to Oberhettinger [66], Papoulis [71, pp. 329–331], Stepanishen [86], and Harris [40] for complete details on the derivations. Next, the notion of an approximate Fourier transform is introduced, and the two diffraction formulations are unified by rigorously demonstrating their Fourier equivalence as an approximate Fourier transform pair. This approximate Fourier equivalence is shown for both unfocused and focused piston transducers. The unfocused Lommel diffraction formulation is discussed first.

1.   THE LOMMEL FORMULATION

Assuming monochromatic and spatially uniform excitation of the unfocused piston transducer shown in Fig. 1.4, $H_1(\rho, z, \omega)$ the scalar disturbance sensed by a fictitious point receiver located at some off-axis distance $\rho = \sqrt{x^2 + y^2}$ can be written

$$H_1(\rho, z, \omega) = \frac{1}{2\pi} \int_{\sigma_o} f(\sigma_o) \frac{e^{-jkr}}{r} \, d\sigma_o \qquad (3.1)$$

where σ_o is the area of the transmitter (aperture) and r is the distance from an elemental area on the face of the transmitter to the point ρ. The subscript o denotes the source ($z = 0$) plane, while the subscript 1 denotes one-way propagation. The velocity distribution across the face of the transducer is $f(\sigma_o)$ which, in our case, is constant due to the assumption of spatial uniformity. For simplicity, we assume $f(\sigma_o) = 1$.

The disturbance $H_1(\rho, z, \omega)$ is known in the literature on ultrasound as the *velocity-potential transfer function* [72]. The meaning of the term will be made clear in the remainder of the chapter. For now, it is sufficient to note that the term *velocity-potential transfer function* implies the existence of a velocity-potential impulse response [86]. The fundamental relationship between an impulse response and its transfer function as an exact Fourier transform pair is well known, and the relevance of this relationship to this chapter shall also be made clear.

Eq. 3.1 represents the Rayleigh-Sommerfeld diffraction integral with an obliquity factor of unity and is applicable to a infinitely baffled flat-faced transducer of any geometry [36, 50]. It is important to note that Eq. 3.1 is based on Hyugen's principle and represents continuous integration of the free-space Green's function for a point source over a continuum which contains, mathematically speaking, an infinite number of point sources.

Per convention, the time dependence of $H_1(\rho, z, \omega)$ on $e^{j\omega t}$ is implied. The spatial wave number $k = 2\pi/\lambda$ is related to temporal frequency ω via $k = \omega/c$. Thus, the dependence of velocity-potential transfer function $H_1(\rho, z, \omega)$ on ω is implicit in two ways.

The Fresnel approximation in conjunction with the circular symmetry of a piston transducer allows the velocity-potential transfer function $H_1(\rho, z, \omega)$ in Eq. 3.1 to be estimated

$$\widehat{H}_1(\rho, z, \omega) = \frac{1}{z} e^{-jk(z+\frac{\rho^2}{2z})} \int_0^a e^{-jk\frac{\rho_o^2}{2z}} J_o\left(\frac{k\rho}{z}\rho_o\right) \rho_o \, d\rho_o, \qquad (3.2)$$

where $\rho_o = \sqrt{x_o^2 + y_o^2}$ is the off-axis distance at the source plane and $\rho = \sqrt{x^2 + y^2}$ is the off-axis distance at the observation plane [71, p. 330]. The hat notation (e.g., $\widehat{H}_1$) indicates that the expression or function is an estimate. Note that $\widehat{H}_1(\rho, z, \omega)$ in Eq. 3.2 is closely related to Φ_P in Reference [81] and is $g(\rho, z)$ in Reference [71, Eq. (3-52),

p. 330] multiplied by j/k. Eq. 3.2 is the classic integral form of Fresnel diffraction from a circular aperture.

A prominent and familiar feature of Fresnel diffraction is its interpretation as a convolution involving a quadratic phase term [34, 71]. This feature is not obvious in Eq. 3.2. However, if the singularity function

$$p_a(\rho_o) = \begin{cases} 1, & \rho_o \le a; \\ 0, & \rho_o > a \end{cases} \tag{3.3}$$

is introduced in the integrand of Eq. 3.2 and the upper limit of integration changed to ∞, then Eq. 3.2 becomes

$$\widehat{H}_1(\rho, z, \omega) = \frac{1}{z}\, e^{-jk(z+\frac{\rho^2}{2z})} \int_0^\infty p_a(\rho_o) e^{-jk\frac{\rho_o^2}{2z}} J_0\left(\frac{k\rho}{z}\rho_o\right) \rho_o\, d\rho_o. \tag{3.4}$$

Eq. 3.4 may be interpreted as the Hankel transform of the product of the singularity function $p_a(\rho_o)$ and a quadratic phase term. The convolution theorem for Hankel transforms allows Eq. 3.4 to be rewritten

$$\widehat{H}_1(\rho, z, \omega) = \frac{1}{k}\, e^{-jk(z+\frac{\rho^2}{2z})} \left[\frac{a}{\rho} J_1\left(\frac{ka\rho}{z}\right) * \frac{1}{j}\, e^{j\frac{k\rho^2}{2z}} \right] \tag{3.5}$$

where the convolution is with respect to $k\rho/z$. The familiar interpretation of Fresnel diffraction is made explicit in Eq. 3.5.

Returning to Eq. 3.2, we note that it can be integrated numerically, but a closed-form expression would simplify matters. A closed-form results if Lommel functions are used. The closed-form result is

$$\widehat{H}_1(\rho, z, \omega) = \frac{1}{k}\, e^{-j(kz+\frac{v^2}{2u}+\frac{u}{2})} \left[U_1(u, v) + jU_2(u, v)\right], \tag{3.6}$$

where the substitutions $u = ka^2/z$ and $v = ka\rho/z$ result in more compact notation. The Lommel functions of two variables, $U_n(u, v)$, were defined in Eq. 1.2 of Chapter 1. Eq. 3.6 is the Lommel diffraction formulation for an *unfocused* piston transducer.

2.   DISCUSSION OF THE LOMMEL DIFFRACTION FORMULATION

In this section, we examine a special case of the Lommel diffraction formulation, revisit work done by Seki, *at al.* [81], and discuss three computational issues associated with the Lommel diffraction formulation. The special case of interest is when $\rho = 0$. In this case, $v = 0$ and $\widehat{H}_1(0, z, \omega)$ describes the on-axis fluctuations of the Lommel diffraction formulation.

Since $U_1(u,0) = \sin(u/2)$ and $U_2(u,0) = \cos(u/2) - 1$ [90, p. 540], it is a matter of simple algebra to show that

$$\widehat{H}_1(0, z, \omega) = \frac{2}{k} e^{-j(kz + + \frac{u}{4})} \sin\frac{u}{4}. \tag{3.7}$$

Thus, the on-axis intensity of the Lommel diffraction formulation is

$$\left|\widehat{H}_1(0, z, \omega)\right|^2 = \frac{4}{k^2} \sin^2\frac{u}{4}, \tag{3.8}$$

which is consistent with the classic result used to define the near field and far field of a circular aperture.

Eq. 3.8 is classically derived by evaluating the integral in Eq. 3.2 with ρ set to zero and subsequently squaring the magnitude of the result. The special case when $\rho = a$ or along the boundary of the geometric shadow of the transducer is left to the reader as an exercise. The interested reader is referred to Papoulis for details [71, p. 331] and is encouraged to compare his or her results with those derived by Williams [95, 96].

Seki, *et al.* used a variant of Eq. 3.6 to calculate pressure as a function of depth z and off-axis distance ρ. The pressure p is related to the velocity-potential transfer function in the following manner,

$$p(\rho, z, \omega, t) = \pm\varrho\widehat{H}_1(\rho, z, \omega)\frac{\partial e^{\pm j\omega t}}{\partial t}, \tag{3.9}$$

where ϱ is medium density [81]. There is disagreement in the literature on sign convention, and $\pm$ in Eq. 3.9 captures this disagreement. The positive convention was chosen in this work. Thus, pressure can be obtained by multiplying Eq. 3.6 by $j\omega\varrho e^{j\omega t}$ [96, p. 286]. Maximum $(e^{j\omega t} = 1)$ pressure responses obtained from Eq. 3.6 are plotted in terms of magnitude and phase in Fig. 3.1; the plots agree well with those in Seki's 1956 paper.

Three computational issues require discussion. First, computation of $U_n(u, v)$ may fail when $u/v > 1$ because the expression converges too slowly in this case. Therefore, it is prudent to compute $U_n(u, v)$ in terms of $V_n(u, v)$ via Eq. 1.4 when $u/v > 1$. Second, the Lommel functions must be determined with a sufficient number of terms to obtain meaningful results; we used $n + 2s \leq 62$ when calculating the Lommel functions. Calculating a Bessel function with this high an order may cause underflow on some computers. Third, on-axis ($\rho = 0$) values of Eq. 3.6 can be calculated via appropriate algorithmic handling of the Lommel functions when $v = 0$, or they may be calculated directly from Eq. 3.7. The latter method was used in our computations.

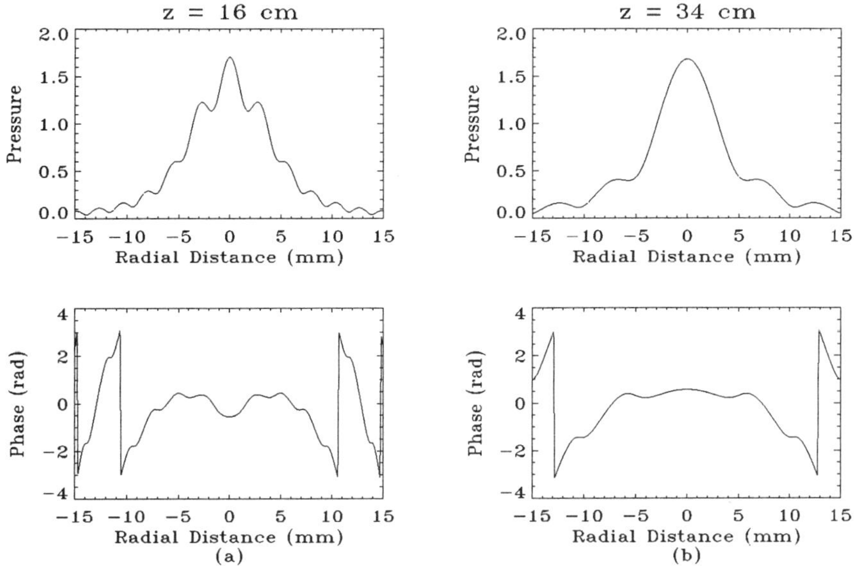

Figure 3.1.   Pressures after Seki, *et al.* [81] via Eq. 3.6

3.   THE ARCCOS FORMULATION

The arccos diffraction formulation can be derived either analytically [66] or geometrically [86]. A modified version of Oberhettinger's analytical derivation [66] is outlined here because it leads to the realization that the arccos and Lommel diffraction formulations form an approximate Fourier transform pair. Assuming the unfocused transducer in Fig. 1.4 is excited by an impulse, the velocity-potential impulse response $h_1(\rho, z, t)$ associated with the fictitious point receiver located at ρ can be written

$$h_1(\rho, z, t) = \frac{1}{2\pi} \int_{-\infty}^{\infty} H_1(\rho, z, \omega) \ e^{j\omega t} \ d\omega, \qquad (3.10)$$
$$= \mathcal{F}^{-1}\{H_1(\rho, z, \omega)\}$$

where $\mathcal{F}^{-1}$ is the inverse Fourier transform. Thus, the velocity-potential impulse response in Eq. 3.10 is the inverse Fourier transform of velocity-potential transfer function $H_1(\rho, z, \omega)$ in Eq. 3.1. The term *velocity-potential impulse response* comes from the fact that the pressure $p(\rho, z, t)$ can be written in terms of a convolution of the piston velocity $u(t)$ with the velocity potential impulse response $h_1(\rho, z, t)$,

$$p(\rho, z, t) = \varrho \frac{\partial}{\partial t} u(t) \underset{t}{*} h_1(\rho, z, t), \tag{3.11}$$

where the convolution is, as indicated, with respect to time [86]. The meaning and significance of the terms *velocity-potential impulse response* and *velocity-potential transfer function* should now be clear.

With a transformation from rectangular to cylindrical coordinates— $x = \rho \cos(\phi)$ and $y = \rho \sin(\phi)$—the velocity potential transfer function $H_1(\rho, z, \omega)$ in Eq. 3.1 becomes

$$H_1(\rho, z, \omega) = \int_{\phi_o=0}^{2\pi} \int_{\rho_o=0}^{a} \left\{ \left[\rho^2 + \rho_o^2 - 2\rho\rho_o \cos(\phi - \phi_o) + z^2 \right]^{-1/2} \times \right.$$

$$\left. \exp\left[-jk\left(\rho^2 + \rho_o^2 - 2\rho\rho_o \cos(\phi - \phi_o) + z^2 \right) \right] \right\} \rho_o \, d\rho_o \, d\phi_o. \tag{3.12}$$

Eq. 3.12 is essentially Eq. 7 in Reference [66]. After several steps, Oberhettinger obtains exact expressions for $h_1(\rho, z, t)$ for two intervals. For $\rho < a$,

$$h_1(\rho, z, t) = \begin{cases} 0, & ct < z; \\ c, & z < ct < R'; \\ \dfrac{c}{\pi} \arccos\left[\dfrac{(ct)^2 - z^2 + \rho^2 - a^2}{2\rho(\,(ct)^2 - z^2\,)^{1/2}} \right], & R' < ct < R; \\ 0, & ct > R, \end{cases} \tag{3.13}$$

and for $\rho > a$,

$$h_1(\rho, z, t) = \begin{cases} 0, & ct < R'; \\ \dfrac{c}{\pi} \arccos\left[\dfrac{(ct)^2 - z^2 + \rho^2 - a^2}{2\rho(\,(ct)^2 - z^2)^{1/2}} \right], & R' < ct < R; \\ 0, & ct > R, \end{cases} \tag{3.14}$$

where $R' = \sqrt{z^2 + (a - \rho)^2}$ and $R = \sqrt{z^2 + (a + \rho)^2}$. Taken together, Eqs. 3.13 and 3.14 represent the arccos diffraction formulation. Methods of dealing with the arccos diffraction formulation at the spatial singularity $\rho = a$ can be found in the literature [14, 40, 86].

Note that Eq. 3.12 is the Rayleigh-Sommerfeld integral for diffraction from a piston transducer, and it led to the arccos diffraction formulation. On the other hand, Eq. 3.2 is the Fresnel approximation to Eq. 3.12, and Eq. 3.2 led to the Lommel diffraction formulation. Clearly, the arccos and Lommel diffraction formulations are closely related, and we will explore this relationship more fully later in the chapter.

4. DISCUSSION OF THE ARCCOS DIFFRACTION FORMULATION

For now, we do well to describe the well-documented behavior of the arccos diffraction formulation [86]. Fig. 3.2, which will be discussed in detail later, can be used as a visual aid. For a fixed depth z, the on-axis velocity-potential impulse response $h_1(\rho = 0, z, t)$ is a rectangular pulse starting at $t = z/c$; its amplitude is c. As ρ increases, the start time of the pulse remains $t = z/c$, but the trailing edge of the pulse moves closer to $t = z/c$. Simultaneously, the fall time of the trailing edge increases, and the trajectory of the fall is governed by the arccos term in Eq. 3.13. In short, the pulse-like nature of the impulse response gradually decays with increasing ρ.

For $\rho > a$, the impulse response no longer resembles a rectangular pulse, and its maximum value is something less than c. In addition, its start time is delayed, and the delay is function of ρ. In the frequency domain then, the high-frequency content of the velocity-potential impulse response from an infinitely baffled piston transducer is maximum on-axis. The high-frequency content decreases with off-axis distance.

For a fixed off-axis distance ρ, the velocity potential impulse response $h_1(\rho, z, t)$ has the same general shape at any depth z but is compressed in time as z increases. The relationship can be quantified by expanding R' and R via binomial expansion, subtracting the smaller from the larger, and dividing for different values of z. The result is that $h_1(\rho, z, t) = h_1(\rho, z, z_r t/z)$ for large z, where z_r is some appropriately chosen reference plane [14]. Researchers in wavelet theory might find this an interesting physical problem since time scaling arises in a natural fashion. Indeed, we shall see that time scaling also occurs in spatially averaged one-way and two-way diffraction.

5. SIMILARITIES AND DIFFERENCES

At this point, the Lommel and arccos diffraction formulations may be compared. The first section of this chapter showed that the Lommel diffraction formulation is a monochromatic frequency-domain expression based on the Fresnel approximation to the Rayleigh-Sommerfeld integral of scalar diffraction theory. Hence, the derivation of Lommel diffraction formulation permits monochromatic diffraction from a circular aperture to be interpreted as a convolution involving a depth-dependent quadratic phase factor (Eq. 3.5). On the other hand, the third section of this chapter showed that the arccos diffraction formulation is a set of polychromatic time-domain expressions based on the exact Rayleigh-Sommerfeld integral (with obliquity factor of unity). The arccos formulation permits impulsive diffraction from a circular aperture to be interpreted in terms of a depth-dependent time-scaling operation. Thus, the two formulations are similar in that they both describe diffraction from a circular aperture, but they differ in derivation (Fresnel vs. Rayleigh-Sommerfeld), realization (frequency vs. time domain) and interpretation (quadratic phase vs. time scaling).

As just mentioned, the Lommel diffraction formulation is based on the Fresnel approximation to the Rayleigh-Sommerfeld diffraction integral. The Fresnel region is often confused with the near field [34], and this confusion may lead to misinterpretation of our results. The following observations are made to avoid confusion and misinterpretation [34, 36, 71].

The Rayleigh-Sommerfeld region consists of the entire half-space in front of the transducer (aperture) [34]. The Fresnel region is that portion of the Rayleigh-Sommerfeld region in which the Fresnel approximation holds [34, 71]. For a circular aperture, this region is delimited by

$$25(a + \rho)^4 < \lambda z^3 = cz^3/f, \qquad (3.15)$$

where λ is wavelength and f is temporal frequency. The requirement in Eq. 3.15 is overly stringent, and it can be shown that the Fresnel approximation is in fact valid for points nearer the transducer (aperture) [36, 71, 99].

The Fraunhofer region is that portion of the Fresnel region where the Fraunhofer approximation holds, and the Fresnel region contains the Fraunhofer region [34, 71]. We follow Robinson, *et al.* [78] and define the near field as the region where the velocity potential oscillates and the far field as the region where the velocity potential decreases monotonically. It is important to note that this definition *is not* based on a point receiver. We also follow Gaskill and emphatically disagree with those who (i) consider the Fresnel and Fraunhofer regions mutually exclusive and (ii) simultaneously equate these regions with the near field and far field, respectively.

The upshot of this discussion is that the Lommel diffraction formulation and expressions based on it are valid in the Fresnel region, the Fraunhofer region, a good portion of the near field, and all of the far field. Consider a brief mathematical argument. In the far field, z is large and $u = ka^2/z$ and $v = ka\rho/z$ in Eq. 3.6 are small for geometries of practical interest. Because the arguments in the Lommel functions are small, terms involving Bessel functions of order greater than one can be neglected. Thus,

$$\widehat{H}_1(\rho, z, \omega) = \frac{a}{k\rho}\, e^{-jk(z+\frac{\rho^2}{2z}+\frac{a^2}{2z})}\, J_1\left(\frac{ka\rho}{z}\right) \qquad (3.16)$$

for large z. Basic trigonometry allows us to write $\rho^2 + z^2 = r^2$, where r, in this case, is the distance from the center of the aperture to the off-axis point ρ. With z large, $\rho/z \approx \sin\theta$ and

$$\widehat{H}_1(\rho, z, \omega) = \frac{a^2}{z}\, e^{-jk(z+\frac{\rho^2}{2z}+\frac{a^2}{2z})}\, \frac{J_1(ka\sin\theta)}{ka\sin\theta}. \qquad (3.17)$$

Eq. 3.17 captures both the far-field directivity pattern of a circular aperture [50] and the celebrated Airy disk pattern [71].

We have just shown that the familiar far-field directivity pattern of a circular aperture can be derived from the Lommel diffraction formulation *as a special case* when z is large. Recall however that the Lommel diffraction formulation is based on the Fresnel approximation and, as a result, holds in the Fresnel region. Thus, we may conclude that the Lommel diffraction formulation and expressions based on it are valid over a large portion of the half space in front of the transducer.

One final point concerning the near field and far field must be made. The conventional notion of a near field and far field is predicated on *monochromatic* excitation. Specifically, the separation between the near field and far field at a given temporal frequency is Z, the last axial maximum as sensed by a fictitious *on-axis point* receiver. For a piston transducer of radius a , the last axial maximum is theoretically located at $Z = a^2/\lambda$ [99]. Recall $c = \lambda f$. Thus, if the excitation is pulsed, the notion of a near field and far field becomes complicated because each frequency in the pulse has an associated Z. This notion is further complicated when spatial averaging is considered. For the purposes of this work, Z is simply the distance at which the *spatially averaged* velocity potential, computed at the center frequency of the transducer, reaches it last axial maximum and thereafter begins to decrease monotonically.

6.   AN APPROXIMATE FOURIER TRANSFORM PAIR

The previous section, taken at face value, leads us to believe that the arccos and Lommel diffraction formulations are, for the most part, quite different. But when placed in the context of Fourier theory, the arccos and Lommel diffraction formulations are more similar than different. Specifically, their differences in terms of realization (time vs. frequency domain) and interpretation (quadratic phase vs. time scaling) become similarities in the context of Fourier theory. Furthermore, the derivation of the arccos diffraction formulation pointed to a close relationship between the arccos and Lommel diffraction formulations. Ultimately, their realization in conjugate domains (time vs. frequency) indicates a possible relationship as an exact Fourier transform pair. This possible Fourier relationship is made more probable by the fact that the interpretation (time scaling vs. quadratic phase) of the two formulations is related to the Fourier explanation of time-scaling as convolution involving quadratic phase terms [71].

Although the derivations of the arccos and Lommel diffraction formulations point to a close relationship between the two formulations, their derivations (Rayleigh-Sommerfeld vs. Fresnel) are quite different. Thus, they cannot form an exact Fourier transform pair. Despite this, they may form an *approximate* Fourier transform pair. This section introduces the notion of an approximate Fourier transform pair and rigorously demonstrates the Fourier equivalence of the arccos and Lommel diffraction formulations as an approximate Fourier transform pair.

Consider a function $f(t)$ that has an exact Fourier transform $F(\omega)$. Mathematically,

$$\mathcal{F}\{f(t)\} = F(\omega) \qquad \text{and} \qquad f(t) = \mathcal{F}^{-1}\{F(w)\}, \qquad (3.18)$$

where $\mathcal{F}$ and $\mathcal{F}^{-1}$ represent the Fourier transform and inverse Fourier transform operations, respectively. Thus, $f(t)$ and $F(\omega)$ form a Fourier transform pair *exactly*. The notion of an *exact* Fourier time-frequency pair has been recognized in the literature on acoustics and ultrasound for decades [64, 72].

The notion of approximate Fourier transform pair is helpful when derivation of an exact Fourier transform pair is too difficult or when it is sufficient to have a rough idea of the Fourier relationship between $f(t)$ and $F(w)$. Mathematically,

$$\mathcal{F}\{f(t)\} \approx \widehat{F}(\omega) \qquad \text{and} \qquad \hat{f}(t) \approx \mathcal{F}^{-1}\left\{\widehat{F}(w)\right\}, \qquad (3.19)$$

where $\hat{f}(t)$ and $\widehat{F}(\omega)$ are estimates of $f(t)$ and $F(\omega)$, respectively. Thus, $f(t) \approx \hat{f}(t)$, and $f(t)$ and $\widehat{F}(\omega)$ form a Fourier transform pair *approximately*.

Consider again Eq. 3.10, the general form of the arccos diffraction formulation. In this equation, the velocity-potential impulse response $h_1(\rho, z, t)$ is the inverse Fourier transform of the velocity-potential transfer function $H_1(\rho, z, \omega)$, and the exact form of $H_1(\rho, z, \omega)$ is unknown. However, a closed-form estimate or approximation is known, namely

$\widehat{H}_1(\rho, z, \omega)$. Thus, we may write

$$\hat{h}_1(\rho, z, t) = \mathcal{F}^{-1}\left\{\widehat{H}_1(\rho, z, \omega)\right\}. \tag{3.20}$$

where $\hat{h}_1(\rho, z, t)$ is an estimate of the impulse response predicted by the arccos diffraction formulation. In short, we claim that the arccos and Lommel diffraction formulations form an approximate Fourier transform pair:

$$\mathcal{F}\{h_1(\rho, z, t)\} \approx \widehat{H}_1(\rho, z, w). \tag{3.21}$$

The claim is verified numerically in the next section.

At this point, a discussion of Gibb's phenomenon and its impact on this work is required [101]. Under the assumptions stated in Chapter 1 and for practical geometries, impulse responses computed using the arccos diffraction formulation have compact support in the time domain; consequently, their Fourier transforms have infinite bandwidth in the frequency domain. In practice, the Lommel diffraction formulation can be sampled only over some finite bandwidth; consequently, impulse responses based on the Lommel diffraction formulation will suffer from Gibb's phenomenon.

As a result, we expect that Lommel-based results will fail to capture temporal discontinuities and will simultaneously exhibit ringing in the neighborhood of any temporal discontinuities. The degree of failure and extent of ringing are functions of the sampling rate; higher sampling rates will capture temporal discontinuities more faithfully but simultaneously introduce more ringing. In short, impulse responses based on the Lommel diffraction formulation and Eq. 3.20 can never show exact agreement with those based on the arccos diffraction formulation in Eqs. 3.13 and 3.14.

The complication just discussed is analogous to the complication encountered in filter design where the desired magnitude-phase response is required to have a discontinuity in the frequency domain [101]. In this case, the desired impulse response has infinite temporal duration. In practice, however, the filter can be sampled only over some finite time duration. Consequently, the realizable filter will exhibit Gibb's phenomenon in the time domain. Filter designers resort to windowing to reduce the effects of Gibb's phenomenon. We will do the same when necessary.

7.   VERIFICATION

The Fourier equivalence of the arccos and Lommel diffraction formulations as an approximate Fourier transform pair will be numerically verified by (i) computing discrete Fourier coefficients using Eq. 3.6, (ii) inverse Fourier transforming these coefficients, and (iii) comparing the results against results obtained from the arccos diffraction formulation in Eqs. 3.13 and Eq. 3.14. Computing discrete Fourier coefficients using Eq. 3.6 is justified by the fact that $k = \omega/c$.

The Lommel diffraction formulation (Eq. 3.6) was used in conjunction with Eq. 3.20 to compute an estimate of the velocity potential impulse response, $\hat{h}_1(\rho, z, t)$, from a piston transducer for three off-axis positions at two depths: $z = 3$ cm and $z = 9$ cm. The speed of sound was set at $c = 1540$ m/s, and the diameter of the piston was set at $2a = 13$ mm. We reiterate that the transducer was assumed to have an infinitely broadband or Dirac response, and the excitation was assumed to be an impulse. The sampling frequency was set at $f_S = 36$ MHz; thus, the Nyquist frequency was 18 MHz.

Note the sampling rate is consistent with 2X oversampling of a real 2.25-MHz piston transducer with a cut-off frequency of 4.5 MHz. More will be said about real transducers at the end of this section. Furthermore, $Z = a^2/\lambda \approx 6$ cm as measured by a point receiver for 2.25-MHz monochromatic excitation. Thus, the xy-planes at $z = 3$ cm and $z = 9$ cm can be considered in the near field and far field, respectively, of a pulsed 2.25-MHz piston transducer with 13 mm diameter.

The results are shown in Fig. 3.2. The off-axis positions are annotated in the figure. The impulse responses for a given ρ are plotted on the same time scale, referenced to $t = z/c$, to emphasize the depth-dependent time scaling discussed previously. In all figures where the two diffraction formulations are compared, Lommel-derived results are plotted with solid lines, while arccos-derived results are plotted with dashed lines. In this work, the arccos diffraction formulation is the gold standard against which the Lommel diffraction formulation is compared.

The plots in Figs. 3.2(a)-(b) show on-axis impulse responses. As was explained earlier, the on-axis impulse response for a piston transducer is a rectangular pulse of amplitude c that gets compressed in time with increasing depth z. The on-axis impulse responses computed with the Lommel diffraction formulation capture this behavior. As a result of Gibb's phenomenon, they do not capture the discontinuities at the beginning and end of each pulse. This was expected.

Since we did not expect exact agreement, we claim that the on-axis impulse responses computed using the Lommel diffraction formulation show satisfactory agreement with the arccos-based results. Figs. 3.2(c)–(d) show impulse responses for $\rho = 3$ mm. With the exception of discontinuities, the Lommel-based results are consistent with the results computed using the arccos diffraction formulation.

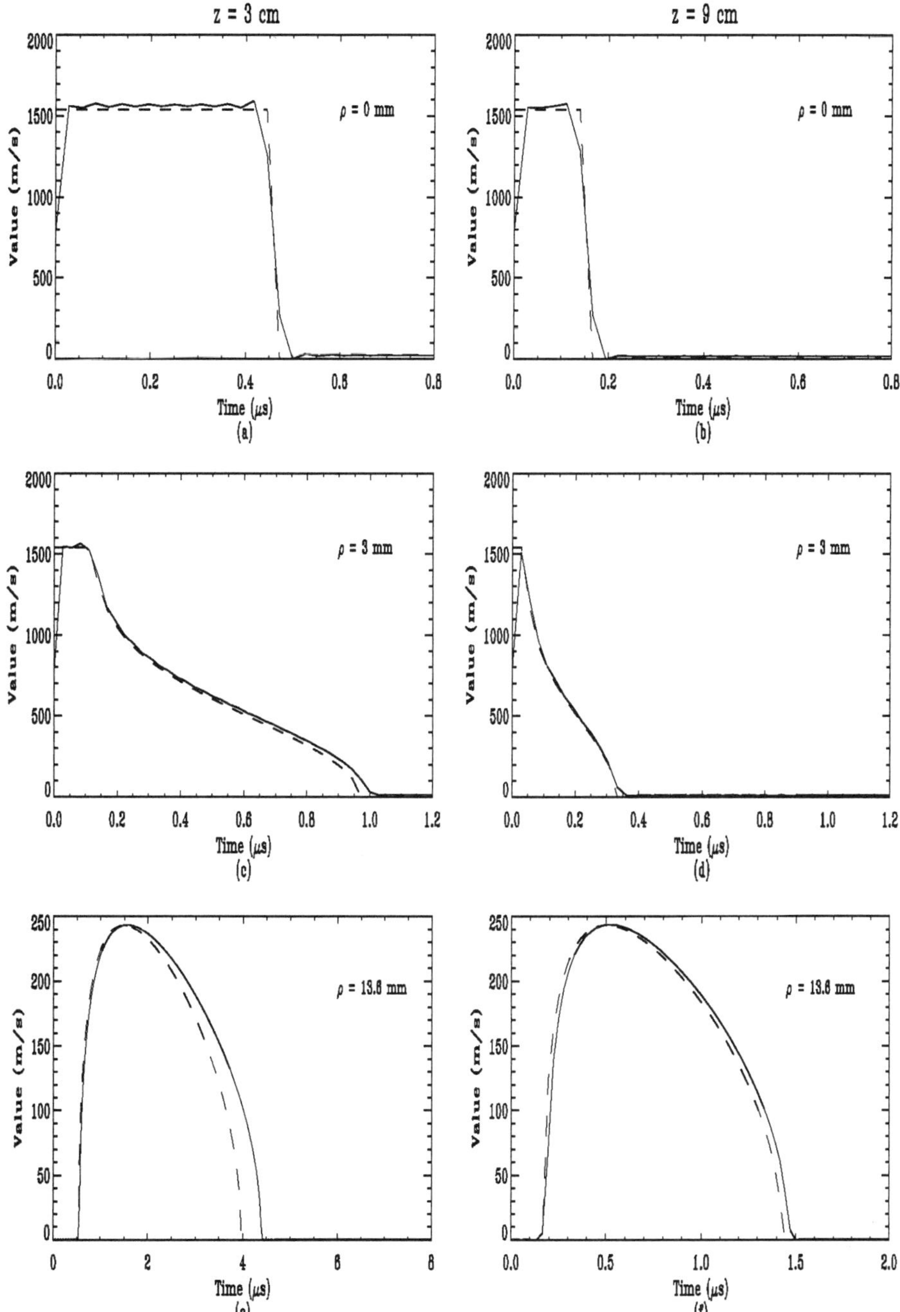

Figure 3.2.   One-way point-receiver impulse responses for the Lommel (solid) and arccos (dashed) diffraction formulations.

The plots in Figs 3.2(e)–(f) show impulse responses for $\rho = 13.6$ mm. Since $\rho > a$, each impulse response should have a maximum amplitude less than c and should start at some time later than $t = z/c$. This behavior is confirmed in the plots. Note that the Lommel diffraction formulation overestimates the time duration of both impulse responses. This is not surprising because the Lommel diffraction formulation is based on the Fresnel approximation which becomes less accurate with increasing off-axis distance ρ and decreasing depth z.

Overall, the results show satisfactory agreement and confirm the validity of the Fourier equivalence of the arccos and Lommel diffraction formulations as an approximate Fourier transform pair. Clearly, the magnitude and phase responses (not shown) computed using the Lommel diffraction formulation capture the salient features of the arccos diffraction formulation. Thus, no discussion of frequency-domain results is included at this point. Frequency-domain results will be discussed in great detail in the chapters that follow.

8.   COMPUTATIONAL CONSIDERATIONS

Three computational issues have been discussed already, and five new ones require discussion. Because they will resurface, these considerations will be referred to as the *five general computational considerations*.

The Fourier transform of a real signal exhibits Hermitian symmetry. Thus, Fourier coefficients need be calculated for positive frequencies only since it is known that the estimated velocity-potential impulse $\hat{h}_1(\rho, z, t)$ is real. Negative-frequency coefficients are computed by simply conjugating the positive ones [62]. This computational benefit is negated by the second computational consideration, namely the fact that the Lommel diffraction formulation is ill-defined at $\omega = 0$. Because of this, a DC frequency coefficient cannot be calculated directly. This consideration is moot if the DC value is not required. If the DC value is required, it can be obtained indirectly by exploiting the fact that the arccos diffraction formulation is positive semi-definite.

Mathematically, $h_1(\rho, z, t) \geq 0$ for all ρ, z, and t of practical interest. In this work, discrete Fourier coefficients were calculated via Eq. 3.6 and inverse Fourier transformed with an FFT algorithm. The resulting samples were forced to be greater than or equal to zero. In short, $\hat{h}_1(\rho, z, t)$ was forced to be positive semi-definite. These two issues represent a trade-off inherent in any Lommel-based solution.

The third issue is the dimensionality of $k = 2\pi f/c$ in the denominator of Eq. 3.6. Since the coefficients calculated from the Lommel formulation are ultimately sent to an FFT algorithm, continuous or discrete frequencies may be used in the computation of k. Discrete frequencies, which are in a sense dimensionless, were used in our implementation. If the estimated impulse response is to be scaled to a maximum value of unity, the choice is immaterial.

Fourth, as explained earlier, estimated impulse responses will suffer from ringing due to Gibb's phenomenon. If desired, this artifact can be reduced with frequency-domain windowing; a window $w(f) = \mathrm{sinc}(0.25\pi f/f_S)$ was used to produce the results shown in Fig. 3.2. The window is admittedly *ad hoc*, but it produced satisfactory results.

Finally, Eq. 3.6 gives no indication of how many frequency samples are required to estimate the arccos impulse response. For a given off-axis position ρ and sampling frequency $f_s = 1/\Delta t$, the minimum number of samples required can be computed via $(R-z)/(c\Delta t)$ or $(R-R')/(c\Delta t)$, whichever is appropriate. Note R and R' are defined in Eqs. 3.13 – 3.14.

A note of caution concludes this section. In numerical comparisons of the two formulations, accurate bookkeeping in terms of phase, zero-padding, sampling frequency, and dimensional scaling is essential because results are being computed in conjugate domains.

9.   THE FOCUSED CASE

The results developed so far apply to unfocused piston transducers only. As can be seen in the literature, focused piston transducers are of great interest to the ultrasound community. For example, O'Neil developed an approximate theory for focused radiators in 1949 [67], and, over a decade later, Kossoff analyzed the focusing action of spherically curved ultrasonic transducers in terms of strong, medium, and weak focusing [52, 53]. In 1974, Penttinen and Luukkala derived a closed-form solution for the velocity-potential impulse response associated with an arbitrary focused piston transducer [72]. Seven years later, Madsen, *et al.* developed a numerical technique for calculating the pressure distribution in the field of a focused piston transducer [61].

The Lommel-based results developed for unfocused piston transducers in the previous sections can be easily extended to focused piston transducers. They can be extended by assuming, as Papoulis [71] did, that focusing introduces a time delay in Eq. 3.6. With this assumption, the Lommel diffraction formulation for a focused piston transducer of radius a is

$$\widehat{H}_1(\rho, z, \omega) = \frac{\epsilon}{kz}\, e^{-j(kz+\frac{k\rho^2}{2z}+\frac{ka^2}{2\epsilon})} \left[U_1\left(\frac{ka^2}{\epsilon}, \frac{ka\rho}{z}\right) + j\, U_2\left(\frac{ka^2}{\epsilon}, \frac{ka\rho}{z}\right)\right],$$

$$(3.22)$$

where $1/\epsilon = 1/z - 1/A$ captures the time delay and A is the focal distance of the transducer [53, 67]. For completeness, we note that strong or short focusing implies $A \approx 0.2a^2/\lambda$, medium focusing implies $A \approx 0.5a^2/\lambda$, and weak or long focusing implies $A \approx 0.8a^2/\lambda$ where a is the radius of the transducer [52, 69].

We will show that Eq. 3.22 is a general linear diffraction formulation for piston transducers because it incorporates all degrees of focusing via the parameter ϵ and includes the unfocused Lommel diffraction formulation as a special case. We will also show that the Lommel diffraction formulation for focused piston transducers in Eq. 3.22 leads directly to the Airy formula when $z = A$ and Kossoff's axial intensity when $\rho = 0$.

In the limit as A approaches infinity, the focal length becomes infinite and the transducer is considered unfocused. Thus, $\lim_{A\to\infty} \epsilon = z$, and Eq. 3.22 becomes Eq. 3.6 which is the Lommel diffraction formulation for unfocused piston transducers. In short, the unfocused Lommel diffraction formulation in Eq. 3.6 is a special case of the focused Lommel diffraction formulation.

At the focus, $z = A$ and when $z = A$, $1/\epsilon = 0$. Thus, it is easy to show with simple algebraic manipulation of the Lommel functions that Eq. 3.22 becomes

$$\widehat{H}_1(\rho, A, \omega) = \frac{a^2}{A} e^{-j(kA + \frac{k\rho^2}{2A})} \frac{J_1(ka\rho/A)}{ka\rho/A}. \tag{3.23}$$

Hence,

$$\left|\widehat{H}_1(\rho, A, \omega)\right|^2 = \left|\frac{a^2}{A} \frac{J_1(ka\rho/A)}{ka\rho/A}\right|^2, \tag{3.24}$$

which is immediately recognized as the Airy formula [67, 71].

It is also a trivial matter to show that the on-axis ($\rho = 0$) fluctuations of the focused Lommel diffraction formulation can be written

$$\widehat{H}_1(0, z, \omega) = \frac{2\epsilon}{kz} e^{-jk(z + a^2/4\epsilon)} \sin\left(\frac{ka^2}{4\epsilon}\right). \tag{3.25}$$

Thus,

$$\left|\widehat{H}_1(0, z, \omega)\right|^2 = \left|\frac{2\epsilon}{kz} \sin\left(\frac{ka^2}{4\epsilon}\right)\right|^2. \tag{3.26}$$

This result is, with the exception of a multiplicative constant, the same result reported by Kossoff for axial intensity [53]. Note that in the unfocused case, $\epsilon = z$, and Eq. 3.25 reduces to Eq. 3.7.

Indirect experimental verification of the focused Lommel diffraction formulation can be had by considering results reported by Madsen, *et al.*

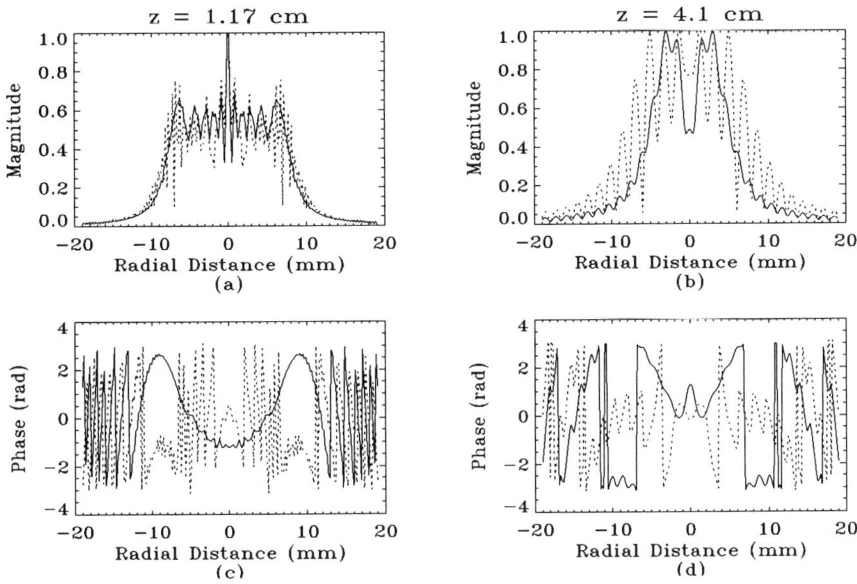

Figure 3.3.   Magnitude and phase responses after Madsen, *et al.* via Eq. 3.22: focused (solid) and unfocused (dotted).

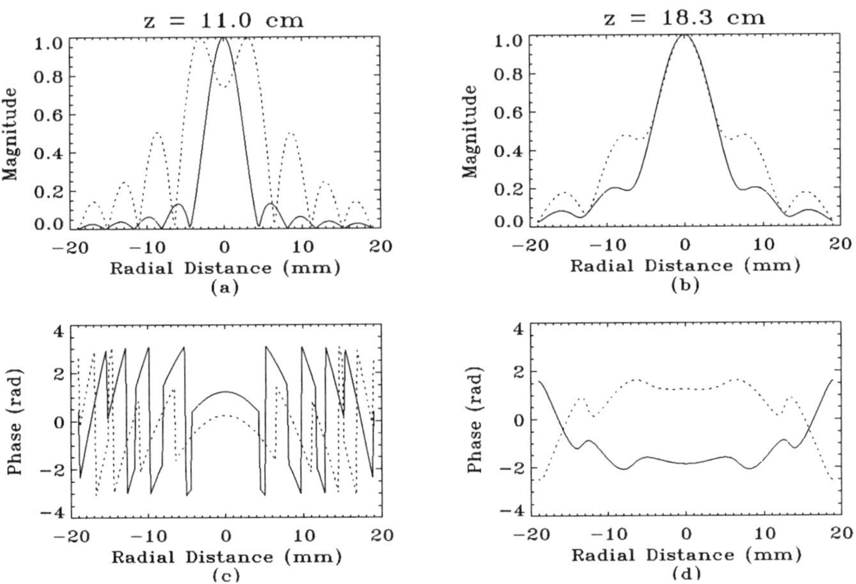

Figure 3.4.   More magnitude and phase responses after Madsen, *et al.* via Eq. 3.22: focused (solid) and unfocused (dotted).

in Reference [61]. Specifically, focused magnitude and phase responses (solid lines) obtained from Eq. 3.22 (multiplied by $-jk$) are plotted in Fig. 3.3 and Fig. 3.4. The following parameters were used in the computations: $2a = 19$ mm, $c = 1540$ m/s, $f = 2.46$ MHz, and $A = 11$ cm [61]. The magnitude and phase responses plotted with dotted lines are for a comparable unfocused transducer and are included as a reference.

The focused magnitude responses (normalized to unity) agree quite well with those in Reference [61]. Indeed, the *theoretical* magnitude responses shown in Figs. 3.3–3.4 are virtually identical to the *experimental* magnitude responses obtained by Madsen and co-workers. On the other hand, the focused phase responses in Figs. 3.3–3.4 exhibit similar features to those in Reference [61], but marked differences are also noticeable. The differences are due to phase wrapping and the fact that Eq. 3.22 is based on the Fresnel approximation.

Despite the differences in phase, the results shown in Figs. 3.3-3.4 serve to validate the use of Eq. 3.22 in the present work. Readers partial to the focused time-domain formalism [72] should note that generating the plots in Figs. 3.3-3.4 presents a substantial computational burden if the time-domain formalism is used. Specifically, a focused impulse response has to be computed for each off-axis position of interest. This would be followed by an FFT operation for each impulse response. In contrast, computation of Eq. 3.22 is relatively easy.

Finally, Eq. 3.22 was used to estimate the velocity-potential impulse response for a focused piston transducer [72] using the principle of approximate Fourier equivalence detailed in the previous two sections. The results are shown in Fig. 3.5. Clearly, the results suffer from Gibb's phenomenon. Overall though, the results show satisfactory qualitative agreement with Fig. 3 in Reference [72]. All computational issues discussed before apply to Eq. 3.22.

10.   CHAPTER SUMMARY

This chapter reviewed derivations of the arccos and Lommel diffraction formulations, two seminal descriptions of diffraction from a circular aperture. The two formulations were compared and shown to be an approximate Fourier transform pair. Their connection was demonstrated numerically for both focused and unfocused piston transducers. Various computational issues were also discussed. The validity and utility of the theoretical insights developed in this chapter are demonstrated in the following chapters.

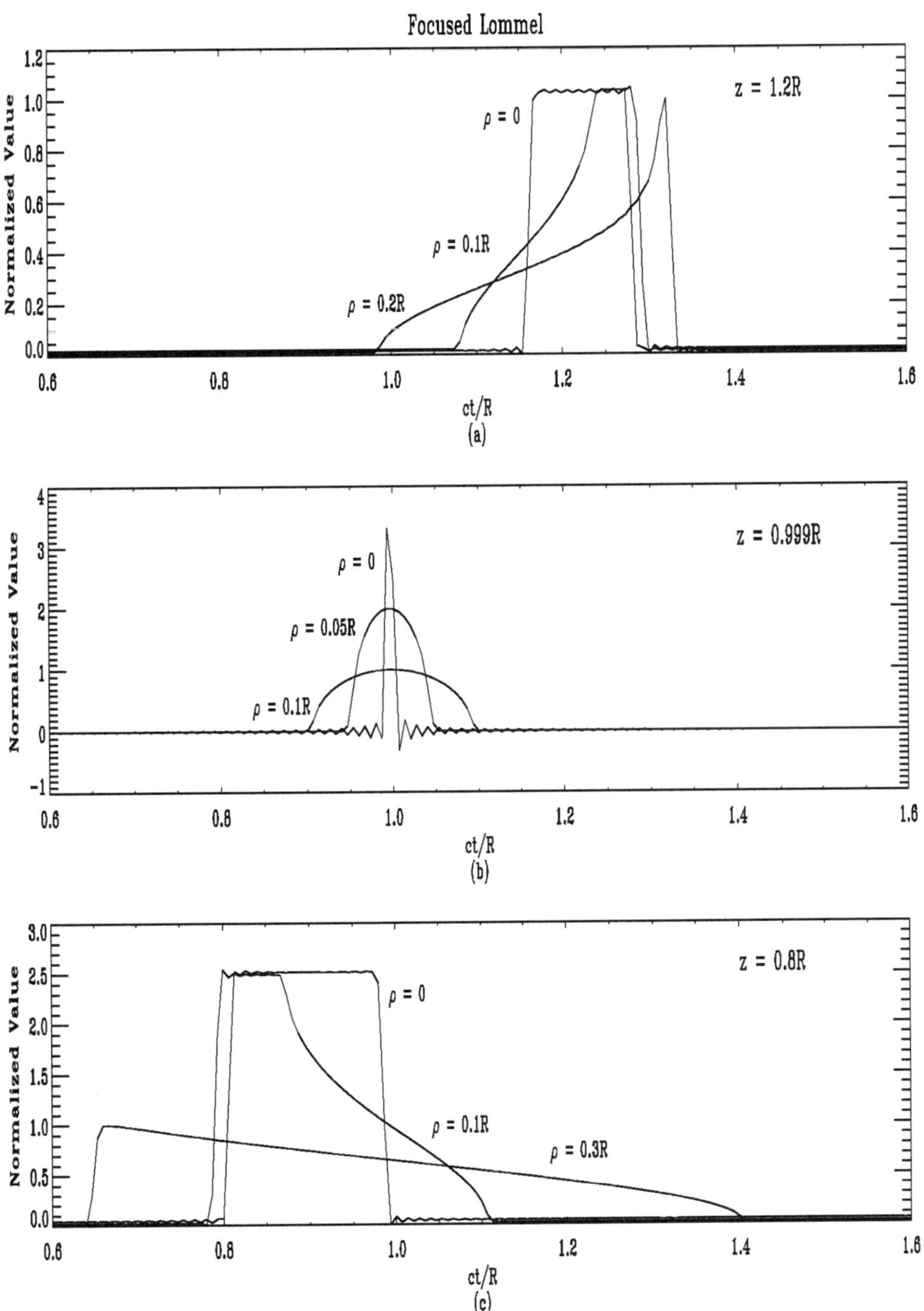

Figure 3.5.   Focused impulse responses after Penttinen and Luukkala [72].

Chapter 4

SPATIALLY AVERAGED ONE-WAY DIFFRACTION

The previous chapter established the Fourier equivalence of the arccos and Lommel diffraction formulations as an approximate Fourier transform pair. In this chapter, we exploit this Fourier equivalence and derive a set of general, closed-form, frequency-domain expressions describing spatially averaged one-way diffraction for *unfocused* piston transmitters and receivers. The expressions derived are general in the sense that the area of the receiver may be less than, equal to, or greater than that of the transmitter. In the time-domain, we present a novel derivation of a closed-form description of one-way diffraction with a finite receiver. Both the time-domain and frequency-domain derivations are followed by discussions and analysis which serve to unify and extend existing theory. Results obtained from the time-domain expressions are compared with those obtained from the frequency-domain expressions. Portions of the material in this chapter first appeared in References [23]-[25] and is reprinted here with permission. Finally, some focused frequency-domain results are derived in Chapter 7.

1. SPATIALLY AVERAGED ARCCOS DIFFRACTION FORMULATION

The velocity-potential impulse response, in integral form, for an unfocused piston transducer and point receiver is

$$
h_1(\rho, z, t) = \begin{cases} ac \int_0^\infty J_0(\tau\rho)\, J_1(\tau a)\, J_0\left(\tau\sqrt{(ct)^2 - z^2}\right)\, d\tau, & ct > z \\ 0, & ct < z \end{cases}
$$

$$(4.1)$$

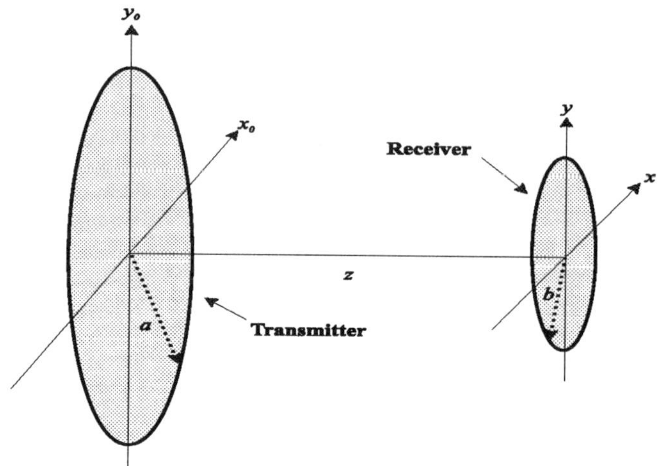

Figure 4.1.   Piston transmitter and finite receiver.

and in closed-form is given by Eqs. 3.13 and 3.14 which involve epony-
mous arccos terms. Eq. 4.1 assumes an unfocused piston transmitter of
radius a and a point receiver as shown in Fig. 1.4.

As explained in the previous chapter, both the integral and closed
forms of the arccos diffraction formulation were derived analytically
by Oberhettinger in 1961 [66]. A decade later, Stepanishen derived
the closed-form solution geometrically and interpreted it as an impulse
response [86]; thus, the arccos diffraction formulation represents the
velocity-potential impulse response for the special case of a piston transducer and *point* receiver.

The spatially averaged impulse response in the case of a *finite* piston
receiver of radius $b \leq a$ coaxially located some distance z in front of the
piston transmitter is

$$\langle h_1(z,t)\rangle_b = \frac{1}{\pi b^2}\left[2\pi\int_0^b h_1(\rho,z,t)\,\rho\,d\rho\right],\qquad(4.2)$$

where $\langle\ \ \rangle_b$ denotes spatial integration and averaging over a disk or
measurement circle of radius b. Note the angular integration over 2π
has already been completed in Eq. 4.2. The situation is depicted in
Fig. 4.1.

Recall the distinction between spatial averaging and spatial integration. The distinction is important because, as was explained in Chapter 1, there is disagreement in the literature on the effects of integration
and averaging. Williams states that the transducer output voltage is

proportional to the spatially integrated pressure impinging on the transducer face aow70, while Harris states that the output voltage is proportional to the spatially averaged pressure [41, p. 187]. The factor of πb^2 is a small but important difference; we will discuss it more later in the chapter. The difference becomes moot if results are normalized to a maximum value of unity.

Cassereau, *et al.* derived a remarkably simple expression for $\langle h_1(z,t)\rangle_b$ valid for any b by integrating the closed-form arccos diffraction formulation directly [14]. Their expression can be obtained with an alternative derivation that provides new insight into diffraction from a circular aperture. The derivation consists of spatially averaging the integral form of the arccos diffraction formulation in Eq. 4.1 and interpreting the result as a Fourier-Bessel or Hankel transform,

$$
\begin{aligned}
\langle h_1(z,t)\rangle_b &= \frac{1}{\pi b^2}\left[2\pi ac \int_{\tau=0}^{\infty}\int_{\rho=0}^{b} J_0(\tau\rho)\,J_1(\tau a)\,J_0\left(\tau\sqrt{(ct)^2 - z^2}\right)\rho\,d\rho d\tau\right],\\
&= \frac{1}{\pi b^2}\left[2\pi abc \int_{\tau=0}^{\infty} \tau^{-1}J_1(\tau b)\,J_1(\tau a)\,J_0\left(\tau\sqrt{(ct)^2 - z^2}\right)d\tau\right],\\
&= \frac{1}{\pi b^2}\left[2\pi abc \int_{\tau=0}^{\infty} \tau^{-2}J_1(\tau b)\,J_1(\tau a)\,J_0(\gamma\tau)\,\tau\,d\tau\right],\\
&= \frac{c}{\pi b^2}\,\mathcal{H}\left\{2\pi\tau^{-1}bJ_1(\tau b)\,\tau^{-1}aJ_1(\tau a)\right\},
\end{aligned}
\tag{4.3}
$$

where $\gamma = \sqrt{(ct)^2 - z^2}$, and $\mathcal{H}$ denotes the Hankel transform with conjugate variables γ and τ.

The convolution theorem for Hankel transforms allows Eq. 4.3 to be written

$$
\langle h_1(z,t)\rangle_b = \frac{c}{\pi b^2}\,\mathrm{cyl}\left(\frac{\gamma}{2b}\right) \underset{\gamma}{*} \mathrm{cyl}\left(\frac{\gamma}{2a}\right),
\tag{4.4}
$$

where $\mathrm{cyl}(\rho/d)$, defined in [34], is a disk of diameter d. Thus, Eq. 4.4 permits the spatially averaged impulse response $\langle h_1(z,t)\rangle_b$ to be interpreted as the convolution of two disks, one with radius a and the other with radius b. This interim result is particularly gratifying given that we have modeled both the transmitter and receiver as disks.

Gaskill and others encountered a similar convolution in their study of optical imaging systems. Gaskill arrived at a closed-form solution via graphical convolution; that is, he calculated the area of overlap of two disks as a function of their center-to-center separation [34, pp. 302–306] as one disk was translated over the other. His solution is directly applicable to the problem at hand.

With $\alpha = (\gamma^2 + a^2 - b^2)/(2\gamma a)$ and $\beta = (\gamma^2 + b^2 - a^2)/(2\gamma b)$, the solution to Eq. 4.4 is

$$\langle h_1(z,t)\rangle_b = \begin{cases} c, & \gamma < a - b; \\ \frac{ca^2}{\pi b^2}\left[\cos^{-1}(\alpha) - \alpha\sqrt{1-\alpha^2}\right] + \\ \quad \frac{c}{\pi}\left[\cos^{-1}(\beta) - \beta\sqrt{1-\beta^2}\right], & a - b \leq \gamma \leq a + b; \\ 0, & \gamma > a + b. \end{cases}$$

$$(4.5)$$

With the exception of a multiplicative constant, Eq. 4.5 is similar to the result derived in Appendix A of Reference [14, Eq. A11], and the interested reader is urged to make a more detailed comparison. The time scaling inherent in the limits of Eq. 3.13 and Eq. 3.14 is also inherent in the limits of Eq. 4.5. That is, $\langle h_1(z,t)\rangle_b$ is compressed in time as z increases.

Though derived under the assumption $b \leq a$, Eq. 4.5 is in fact completely general. When $b > a$, the two variables can simply be interchanged [14]. When $a = b$, Eq. 4.5 can be easily manipulated into the closed-form solution derived by Rhyne [76]. Derivation of closed-form solution for the focused case remains fertile ground for research [2, 72].

It is important to note that although Rhyne's result holds for one-way diffraction, he derived his result by invoking the mirror-interpretation of two-way diffraction involving an infinite plate. Specifically, Rhyne did *not* integrate Eq. 4.2 over an infinite extent. Instead, he integrated Eq. 4.2 with $b = a$ and then invoked the mirror-image interpretation of ultrasonic reflection from an infinite plate.

This discussion leads to two very subtle points. First, the mirror-image interpretation holds only for acoustic coupling to and from an infinitely large and perfect ultrasonic reflector. Second, the mirror-image interpretation permits the application of Eq. 4.5 when $b = a$; otherwise, it does not. We reiterate, however, that Eq. 4.5 is completely general for spatially averaging one-way diffraction effects involving an unfocused piston transmitter and receiver.

2.   ANALYSIS OF TIME-DOMAIN RESULTS

Additional insight can be gained by considering Eqs. 4.4–4.5 in more detail. First, the commutativity of the convolution operation in Eq. 4.4 allows us to claim that Eq. 4.5 is general, and the commutativity of the convolution may be interpreted as a mathematical manifestation of Helmholtz's reciprocity theorem [51]. Second, Eq. 4.5 is well known in optics. Gaskill calls it the *cylinder-function cross correlation* [34,

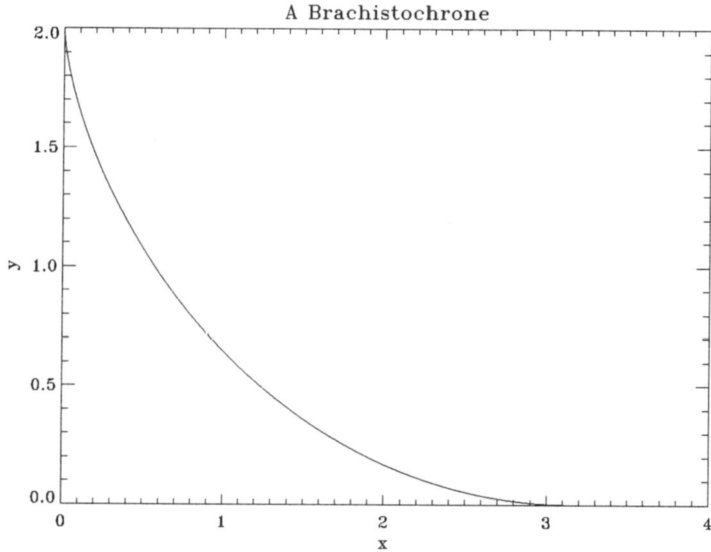

Figure 4.2.   A famous cycloid.

pp. 302–304], while Bracewell calls the special case that results when $b = a$ the *chat* function [11, pp. 187–192]. Bracewell coined this term because the shape of the graph of Eq. 4.5, plotted as a function of γ for $a = b$, resembles a Chinese farmer's hat. The *chat* function characterizes the modulation transfer function for a perfect, diffraction-limited circular lens [56] and is similar in form to Eq. 4.5 when $b = a$. Thus, insights and results developed in optics for Eq. 4.5 may benefit researchers in ultrasound.

Finally, the graph that results from plotting Eq. 4.5 as a function of time t for $b = a$ is strikingly similar to the aesthetically pleasing *brachistochrone* [21, 35, 46]. Compare Fig. 4.2 and Figs. 4.8(a)–(b). Indeed, the form of Eq. 4.5 is similar to the equation for the brachistochrone when $b = a$. Specifically, Eq. 4.5 reduces to

$$\langle h_1(z,t)\rangle_a = \frac{2c}{\pi}\left[\arccos\left(\frac{\gamma}{2a}\right) - \frac{\gamma}{2a}\sqrt{1 - \frac{\gamma^2}{4a^2}}\,\right], \tag{4.6}$$

when $b = a$ and, with the parameterization

$$x = r(\theta - \sin\theta), \tag{4.7a}$$

$$y = r(1 + \cos\theta), \tag{4.7b}$$

the equation for the brachistochrone is

$$x = r \arccos\left(\frac{y-r}{r}\right) - r\sqrt{1 - \left(\frac{y-r}{r}\right)^2}. \qquad (4.8)$$

Note the similarity between Eq. 4.6 and Eq. 4.8. The graphical and functional similarities just noted are not surprising when one considers the physical origin of and mathematical solution to the brachistochrone and diffraction problems. Specifically, both problems can be formulated in terms of Hamilton's principle of least action [35]. Thus, both are mathematically amenable to solution via the calculus of variations. A more rigorous and extensive comparison of these two problems may lead to deeper understanding of diffraction from piston transducers and transducers with more general geometries.

3.   SPATIALLY AVERAGED LOMMEL DIFFRACTION FORMULATION

A set of closed-form *frequency-domain* expressions describing the spatially averaged one-way diffraction effects sensed by a finite receiver can be obtained by spatially averaging the *unfocused* Lommel diffraction formulation in Eq. 3.6. A spatially averaged closed-form expression will be derived for each of three cases: a piston receiver with radius (i) $b < a$, (ii) $b = a$, and (iii) $b > a$. In each case, the transmitter, a piston transducer with radius a, and the receiver are coaxial and separated by a distance z.

Spatial averaging of the Lommel diffraction formulation is facilitated by Eqs. 1.2–1.10, results derived by Wolf [97], and two lemmas found in the appendix of this chapter. Spatially integrating and averaging Eq. 3.6 yields

$$\langle \widehat{H}_1(z,\omega)\rangle_b = \frac{1}{\pi b^2}\left[2\pi \int_0^b \widehat{H}_1(\rho,z,\omega)\rho\,d\rho\right]. \qquad (4.9)$$

Note the angular integration from 0 to 2π has been completed, and the notation is as before.

The integral in Eq. 4.9 can be solved for $b < a$ with the help of results derived by Wolf [97]. Specifically, the $U_n(u,v)$ functions in the integrand are expanded in terms of $V_n(u,v)$ via Eq. 1.4 and integrated using Lemma 9 on p. 548 of Reference [97]. The intermediate result is simplified by noting that $W_1(u,v) - W_3(u,v) = Y_1(u,v)$ and $2W_2(u,v) = Y_2(u,v)$. With $v_b = kab/z$ and $u = ka^2/z$, the first spatially averaged expression is

$$\langle \widehat{H}_1(z,\omega)\rangle_{b<a} = -\frac{2z}{(kb)^2}\, e^{-j(kz+\frac{u}{2})} \quad \times$$

$$\left\{ j\frac{v_b^2}{2u}\, e^{j\frac{u}{2}} + e^{-j\frac{v_b^2}{2u}}\, [\,Y_2(u,v_b) - jY_1(u,v_b)\,] \right\}. \quad (4.10)$$

The Y_n functions are highly convergent and fairly easy, in terms of coding and processor time, to compute.

In the second case, $b = a$, $v_b = u$, and Eq. 4.9 succumbs to a welcome simplification that is a special case of Eq. 4.10. Specifically, $Y_1(u,u) = 0.5uJ_0(u)$ and $Y_2(u,u) = 0.5uJ_1(u)$; thus the expression for spatially averaged one-way diffraction effects when $b = a$ is

$$\langle \widehat{H}_1(z,\omega)\rangle_{b=a} = -\frac{2z}{(kb)^2}\, e^{-j(kz+\frac{u}{2})} \quad \times$$

$$\left\{ j\frac{u}{2}\, e^{j\frac{u}{2}} + e^{-j\frac{u}{2}} \left[\frac{u}{2}J_1(u) - j\frac{u}{2}J_0(u) \right] \right\}. \quad (4.11)$$

Eq. 4.11 is a special case of Eq. 4.10 and is very easy to compute. It may also be applied to the mirror-image interpretation of two-way diffraction involving an infinite plate.

Closed-form results have been derived for the first two cases: $b < a$ and $b = a$. The last case, $b > a$, is a bit more tedious. The result is derived by splitting the range of integration in Eq. 4.9 into two intervals: $0 \le \rho \le a$ and $a \le \rho \le b$. The problem then becomes

$$\langle \widehat{H}_1(z,\omega)\rangle_{b>a} = -\frac{2z}{(kb)^2}\, e^{-j(kz+\frac{u}{2})} \quad \times$$

$$\left\{ j\frac{u}{2}\, e^{j\frac{u}{2}} + e^{-j\frac{u}{2}} \left[\frac{u}{2}J_1(u) - j\frac{u}{2}J_0(u) \right] \right\} \quad +$$

$$\frac{1}{\pi b^2} \left[2\pi \int_a^b \widehat{H}_1(\rho,z,\omega)\rho\, d\rho \right] \quad (4.12)$$

because we have just found the closed-form solution to the integral over the interval $0 \le \rho \le a$. With $u = ka^2/z$, $v = ka\rho/z$, and $v_b = kab/z$, the integral in Eq. 4.12 is

$$\langle \widehat{H}_1(z,\omega)\rangle_{u \le v \le v_b} = \frac{2z}{u(kb)^2}\, e^{-j(kz+\frac{u}{2})} \int_u^{v_b} [\,U_1(u,v) + jU_2(u,v)\,]\, e^{-j\frac{v^2}{2u}}\, v\, dv.$$

$$(4.13)$$

Application of Lemma 2 in the appendix at the end of this chapter and some algebra yields

$$\langle \widehat{H}_1(z,\omega)\rangle_{u\le v\le v_b} = \frac{2z}{(kb)^2} e^{-j(kz+\frac{u}{2})} \times$$

$$\left\{ e^{-j\frac{u}{2}} \left[2Z_2(u,u) - jZ_1(u,u) + jZ_3(u,u)\right] \; + \right.$$

$$\left. e^{-j\frac{v_b^2}{2u}} \left[jZ_1(u,v_b) - jZ_3(u,v_b) - 2Z_2(u,v_b)\right] \right\}, \quad (4.14)$$

which may be simplified because $Z_1(u,v) - Z_3(u,v) = X_1(u,v)$ and $2Z_2(u,v) = X_2(u,v)$. Thus, Eq. 4.14 becomes

$$\langle \widehat{H}_1(z,\omega)\rangle_{u\le v\le v_b} = \frac{2z}{(kb)^2} e^{-j(kz+\frac{u}{2})} \times$$

$$\left\{ e^{-j\frac{u}{2}} \left[X_2(u,u) - jX_1(u,u)\right] \; + \right.$$

$$\left. e^{-j\frac{v_b^2}{2u}} \left[jX_1(u,v_b) - X_2(u,v_b)\right] \right\}. \quad (4.15)$$

But, $X_1(u,u) = 0.5uJ_0(u)$ and $X_2(u,u) = 0.5uJ_1(u)$, so

$$\langle \widehat{H}_1(z,\omega)\rangle_{u\le v\le v_b} = \frac{2z}{(kb)^2} e^{-j(kz+\frac{u}{2})} \times$$

$$\left\{ e^{-j\frac{u}{2}} \left[\frac{u}{2}J_1(u) - j\frac{u}{2}J_0(u)\right] \; + \right.$$

$$\left. e^{-j\frac{v_b^2}{2u}} \left[jX_1(u,v_b) - X_2(u,v_b)\right] \right\}. \quad (4.16)$$

Substitution of Eq. 4.16 into Eq. 4.12 and a little algebra produces the third closed-form frequency-domain expression:

$$\langle \widehat{H}_1(z,\omega)\rangle_{b>a} = -\frac{2z}{(kb)^2} e^{-j(kz+\frac{u}{2})} \times$$

$$\left\{ j\frac{u}{2} e^{j\frac{u}{2}} + e^{-j\frac{v_b^2}{2u}} \left[X_2(u,v_b) - jX_1(u,v_b)\right] \right\}. \quad (4.17)$$

The X_n functions are highly convergent and fairly easy, in terms of coding and processor time, to compute.

The closed-form results derived in this section are compiled below for convenience:

$$\langle \widehat{H}_1(z,\omega)\rangle_b = -\frac{2z}{(kb)^2}\, e^{-j(kz+\frac{u}{2})} \quad \times$$

$$\begin{cases} \left(j\frac{v_b^2}{2u} e^{j\frac{u}{2}} + e^{-j\frac{v_b^2}{2u}} \left[Y_2(u,v_b) - jY_1(u,v_b) \right] \right), & b < a; \\[2ex] \left(j\frac{u}{2} e^{j\frac{u}{2}} + e^{-j\frac{u}{2}} \left[\frac{u}{2} J_1(u) - j\frac{u}{2} J_0(u) \right] \right), & b = a; \\[2ex] \left(j\frac{u}{2} e^{j\frac{u}{2}} + e^{-j\frac{v_b^2}{2u}} \left[X_2(u,v_b) - jX_1(u,v_b) \right] \right), & b > a. \end{cases} \quad (4.18)$$

Observe the similarity and symmetry of the closed-form expressions when $b < a$ and when $b > a$; we will comment on this observation in the next section. For now, the similarity and symmetry are comforting from a mathematical and physical perspective given the nature of the functions involved.

4.   ANALYSIS OF FREQUENCY-DOMAIN RESULTS

Additional insight can be gained by analyzing Eq. 4.10, Eq. 4.11, and Eq. 4.17 in detail. First, we begin with Eq. 4.10 and show that it reduces to Eq. 3.7 in the limit as b approaches zero. Second, we compare Eq. 4.11 to results derived a few decades ago. Third, we let b approach infinity in Eq. 4.17. Finally, we examine the effects of spatially averaging on the velocity-potential transfer function in the near field and far field.

Consider Eq. 4.10 in the limit as b approaches zero. First, note that the leading real factor $2z/(kb)^2$ can be written $2u/(kv_b^2)$. Second, it is a simple matter to show that

$$\lim_{b\to 0} \frac{2u}{v_b^2}\, Y_1(u,v_b) = 1, \qquad (4.19)$$

and that

$$\lim_{b\to 0} \frac{2u}{v_b^2}\, Y_2(u,v_b) = 0, \qquad (4.20)$$

where $u = ka^2/z$ and $v_b = kab/z$ as before. The former equation follows since $\lim_{x\to 0} J_1(x)/x = 1/2$. These two facts and a little algebra allow us to write

$$\lim_{b\to 0}\langle \widehat{H}_1(z,\omega)\rangle_b = \frac{2}{k}\, e^{-j(kz++\frac{u}{4})} \sin\frac{u}{4}. \qquad (4.21)$$

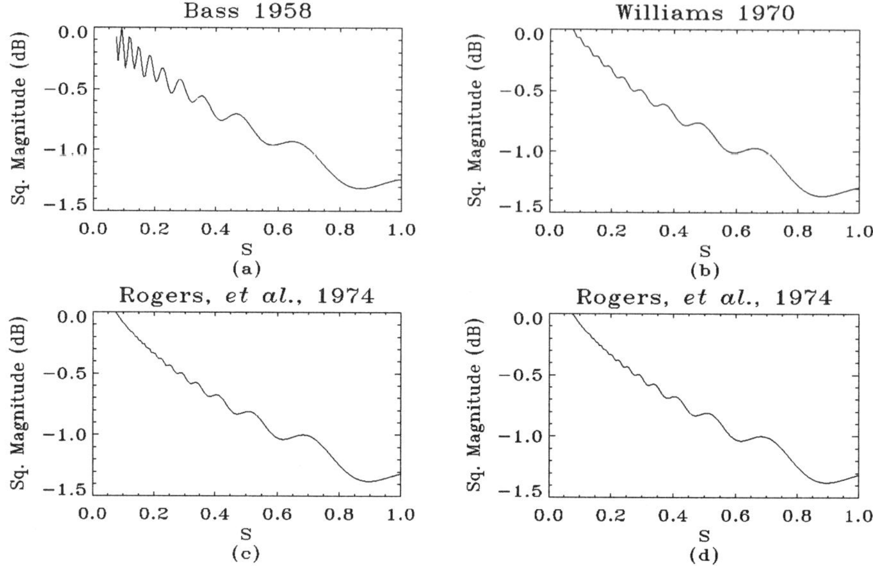

Figure 4.3. Attenuation caused by diffraction as a function of S in the near-field.

which is the same as Eq. 3.7. Thus, the *spatially averaged version* of the Lommel diffraction formulation in Eq. 4.10 incorporates the *on-axis, point-receiver* description of the Lommel diffraction formulation in Eq. 3.7 as a special case.

We now turn our attention to Eq. 4.11. It is, with the exception of a multiplicative constant, the same as the result derived by Rogers and Van Buren [79]. Thus, we compare results obtained from Eq. 4.11 to results derived by Bass [5] and Williams [96]. The results are plotted Figs. 4.3 and 4.4. As an aside , Chen, *et al.* derived a result similar to Eq. 4.11 for the monochromatic pressure transfer function of a *focused* transducer [16].

The parameters used to compute the squared-magnitude results plotted in the figures are the same as were used by Bass: $c = 1200$ m/s, $a = 1$ cm, and $f = 0.956$ MHz. The data were obtained from (i) Bass's 1958 equation [5, Eq. (14)], (ii) Williams' 1970 equation [96, Eq. (6)], and (iii) Eq. 4.11. Note that Williams [96, p. 286] corrected two typos in Bass's 1958 equation. The graphs show the spatially integrated diffraction effects, due to monochromatic excitation, plotted as a function of $S = z\lambda/a^2$. Two graphs of Rogers and Van Buren's result are included for the purposes of comparison.

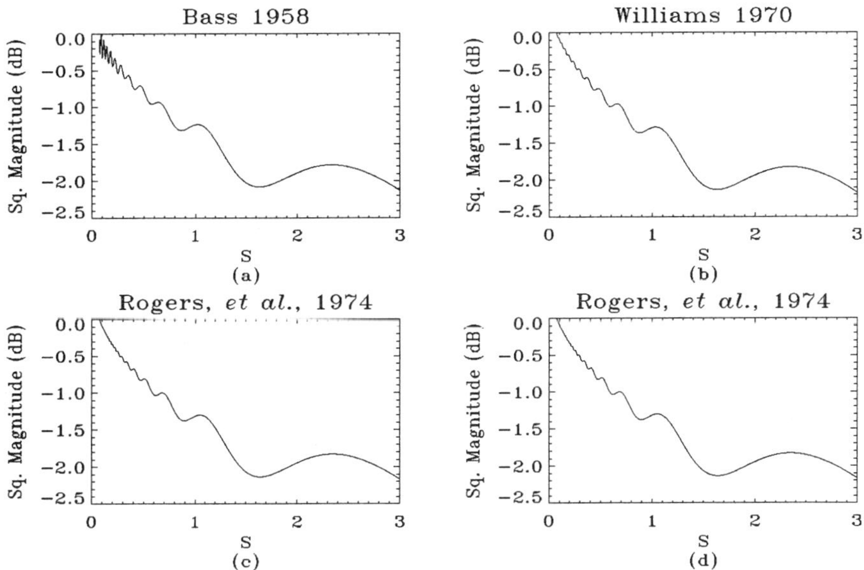

Figure 4.4.   Attenuation caused by diffraction as a function of S in both the near-field and far-field.

The oscillatory behavior of Bass's result at low S is due to small number of terms used in his equation. The overall results, however, show excellent agreement, and the plots confirm the well-established fact that attenuation due to diffraction increases with depth z. Thus, we have gained new academic perspective on classic research with the help of Eq. 4.11.

Interesting insight into the nature of one-way diffraction with an infinite receiver can be gained by letting b go to infinity in Eq. 4.17 and considering the total pressure impinging on the infinite receiver. Multiplying Eq. 4.17 by πb^2 and some further algebra yields

$$(\pi b^2)\,\langle \widehat{H}_1(z,\omega)\rangle_b = -j\frac{\pi a^2}{k}\,e^{-jkz} - 2ze^{-j(kz+\frac{u}{2}+\frac{v_b^2}{2u})} \quad \times$$

$$[X_2(u,v_b) - jX_1(u,v_b)]. \quad (4.22)$$

Letting b approach infinity yields

$$\lim_{b\to\infty} (\pi b^2)\langle \widehat{H}_1(z,\omega)\rangle_b = -j\frac{\pi a^2}{k}e^{-jkz}. \quad (4.23)$$

The same result can be obtained by spatially integrating the unfocused Lommel diffraction formulation of Eq. 3.6 directly. Doing so yields

$$\lim_{b\to\infty} (\pi b^2)\langle \widehat{H}_1(z,\omega)\rangle_b = \lim_{b\to\infty} 2\pi \int_0^b \widehat{H}_1(\rho,z,\omega)\,\rho d\rho \qquad (4.24)$$

where the upper limit of infinity is not problematic because of the rapidly converging Lommel functions in the integrand. Eq. 4.23 can be obtained from Eq. 4.24 with the help of Watson [90, p. 541], Wheelon [94, pp. 76–77, Eq. 1.608 and Eq. 1.610], and Euler's formula.

At any rate, the closed-form result in Eq. 4.23 is virtually identical to that reported by Williams for monochromatic diffraction with a theoretically infinite receiver [96, Eq. 40]. If Eq. 4.23 is used to calculate total maximum pressure, it yields the pressure, $c\varrho\pi a^2 e^{-jkz}$, "produced by a section of area πa^2 cut out of a plane wave that has the same particle velocity, [unity in this case], as does the piston source. [96, p.289]." In essence, the magnitude of the total maximum pressure detected by the infinite receiver is the same at all z-planes. No pressure/energy is lost because (i) the receiver is infinite and (ii) no loss mechanism has been introduced into the theory. In this case, diffraction introduces only a depth-dependent phase shift via the e^{-jkz} term.

Further insight can be gained by examining Eq. 4.22 and Eq. 4.23 more closely. Note that the first term in Eq. 4.22 is identical to right-hand side of Eq. 4.23. Thus, the first term in Eq. 4.22 represents the infinite-receiver solution, and the second term represents the influence of diffraction [5]. Similar observations can be made about Eq. 4.10 and Eq. 4.11. Furthermore, the $k = \omega/c$ in the denominator of the infinite receiver solution of Eq. 4.23 indicates that one-way diffraction as measured by an infinite receiver is dominated by a $1/f$ lowpass filtering effect. This claim will be discussed in more detail later.

Finally, Eq. 4.18 was computed for six cases of the parameter b; the solid-line graphs in Fig. 4.5 show the squared magnitude of the results plotted as a function of $S = z\lambda/a^2$ for each case. The dotted line in each graph is a plot of Eq. 3.8, the intensity of the on-axis Lommel diffraction formulation, normalized to its maximum value and plotted as a function of S. All results shown in Fig. 4.5 are plotted in decibels (dB).

Three observations require reiteration. First, Eq. 3.8 represents the squared magnitude of the Lommel diffraction formulation along the z-axis [71]. Second, $S = 1$ is the conventional demarcation between the near field and far field [34] for a circular aperture when a *point receiver* is a assumed. Third, for the purposes of this work, we designate the distance z at which the *spatially averaged* velocity potential begins to decrease monotonically as Z.

Fig. 4.5(a) shows the spatially averaged diffraction field when $b = a/100$. In this case, the on-axis receiver or measurement circle is essentially a point, and the results should be similar to those obtained from Eq. 3.8. The graph in Fig 4.5(a) indicates the results are virtually identical. Indeed, Z, the position of the last axial maximum, occurs at $S = 1$. This result should not be surprising since we have already shown that the *spatially averaged version* of the Lommel diffraction formulation in Eq. 4.10 incorporates the *on-axis, point-receiver* description of the Lommel diffraction formulation in Eq. 3.7 as a special case.

Fig. 4.5(b) shows the spatially averaged diffraction field when $b = a/5$. The graph indicates that, in the near field at least, the spatially averaged diffraction field (solid line) is markedly different than the on-axis diffraction field (dotted line) predicted by Eq. 3.8. Despite this difference, the far field behavior plotted for each case is quite similar. Note, in particular, that Z for each case occurs at the same value of S.

Fig. 4.5(c) shows a graph of the spatially averaged diffraction field when $b = a/2$ along with a graph of the on-axis diffraction field. Curiously, Z, the position of last axial maximum, occurs at a different value of S for each case. Fig. 4.5(d) shows the spatially averaged diffraction when $b = a$. The graph is consistent with the literature [79] and shows that the last axial maximum has been averaged out. In this case, there is no clear demarcation between the near field and far field. Finally, Fig. 4.5(e)-(f) confirms the fact that the Fresnel approximation begins to break down when $b > a$.

The results shown in Fig. 4.5 lead to two conclusions. First, the behavior of the spatially averaged diffraction field is virtually identical to that of the on-axis diffraction field, at least in a squared-magnitude sense, for values of b an order of magnitude greater than $b = a/100$. Second, spatial averaging smoothes out velocity-potential oscillations in the near field, decreases the rate of monotonic fall-off in the far field, and eventually averages the near-field out as b is increased. The interested reader may wish to explore how spatial averaging affects phase.

5.   EXTENDING FOURIER EQUIVALENCE

We hypothesize that spatial averaging is, in terms of Fourier theory, the same in both the time domain and the frequency domain. Thus, we should be able to extend the approximate Fourier equivalence of the arccos and Lommel diffraction formulations to spatially integrated one-way diffraction. Spatially averaging Eq. 3.1 and subsequently inverse Fourier transforming the result yields

$$\langle h_1(z,t) \rangle_b = \mathcal{F}^{-1}\left\{ \langle H_1(z,\omega) \rangle_b \right\}. \tag{4.25}$$

Eq. 4.25 is the mathematical statement of our hypothesis. We extend the approximate Fourier equivalence developed in the previous chapter to spatially averaged diffraction and claim that

$$\langle H_1(z,\omega)\rangle_b \approx \langle \widehat{H}_1(z,\omega)\rangle_b. \tag{4.26}$$

As a result, we may write

$$\langle h_1(z,t)\rangle_b \approx \mathcal{F}^{-1}\left\{\langle \widehat{H}_1(z,\omega)\rangle_b\right\}. \tag{4.27}$$

But we also have

$$\langle \hat{h}_1(z,t)\rangle_b = \mathcal{F}^{-1}\left\{\langle \widehat{H}_1(z,\omega)\rangle_b\right\}. \tag{4.28}$$

Thus,

$$\langle h_1(z,t)\rangle_b \approx \langle \hat{h}_1(z,t)\rangle_b. \tag{4.29}$$

Theoretically then, substituting $k = \omega/c$ in Eq. 4.18 should allow estimation of the Fourier coefficients for the spatially averaged arccos impulse response sensed over some measurement circle of radius b. These coefficients can then be inverse Fourier transformed to estimate the spatially averaged arccos impulse response. This reasoning is simply an extension of the approximate Fourier equivalence of the Lommel and arccos diffraction formulations developed in Chapter 3 for a point receiver.

Some discussion is required before computing and comparing spatially averaged impulse responses. First, Eq. 4.5 is a closed-form time-domain expression for the spatially averaged arccos diffraction formulation. It serves as the gold standard in this chapter.

Second, impulse responses computed using Eq. 4.5 have compact support in the time domain; consequently, their Fourier transforms have infinite bandwidth in the frequency domain. Like the Lommel diffraction formulation for a point receiver (Eq. 3.22), Eq. 4.18 must be sampled over some finite bandwidth. Consequently, impulse responses based on Eq. 4.18 will suffer from Gibb's phenomenon, and the comments made about impulse responses based on the Lommel diffraction formulation in Chapter 3 apply.

6.   VERIFICATION

The Fourier equivalence of the arccos and Lommel diffraction formulations as an approximate Fourier transform pair predicts that Eq. 4.18 may be used to estimate the one-way spatially averaged impulse response associated with the arccos diffraction formulation. Eq. 4.18 was used in this fashion, and the results were plotted against the spatially averaged arccos diffraction formulation of Eq. 4.5. Fig. 4.6 through Fig. 4.10 show the plots. Five values of b were used, and they are annotated in the graphs. Only Fig. 4.8 is discussed in detail. Concise comments pertaining to Figs. 4.6–4.7 and Figs. 4.9–4.10 follow the discussion of Fig 4.8.

Figs. 4.8(a)–(b) show spatially averaged one-way impulse responses estimated via Eq. 4.11 (solid lines) and spatially integrated one-way impulse responses calculated by using Eq. 4.5 (dashed lines). Before proceeding, we note again in passing the resemblance of the impulse responses to the brachistochrone shown in Fig. 4.2.

The impulse responses were calculated for $b = a$ at two depths: $z = 3$ cm and $z = 9$ cm. The speed of sound was set at $c = 1540$ m/s, and the diameter of the piston was set at $2a = 13$ mm. The transducer was assumed to have an infinitely broadband response, and the excitation was assumed to be an impulse. The sampling frequency was set at $f_S = 36$ MHz; thus, the Nyquist frequency was 18 MHz.

With the exception of discontinuities, the impulse responses based on the spatially averaged Lommel formulation in Eq. 4.11 are consistent with the results computed using the spatially averaged arccos formulation in Eq. 4.5 and with results computed by Kuc and Regula [55]. They even capture the depth-dependent temporal compression discussed earlier.

The five general computational issues discussed in the previous chapter apply to spatially averaged one-way diffraction. Indeed, the window $w(f)$ described in Chapter 3 was used in computing the spatially averaged Lommel-based impulse responses. Thus, ringing due to Gibb's phenomenon is reduced in the plots, and the impulse responses show satisfactory agreement. It is also important to reiterate the ease with which Eq. 4.11 can be computed.

Figs. 4.8(c)-(d) and Figs. 4.8(e)-(f) show the squared magnitude responses (dB) and the phase responses associated with the impulse responses in Fig. 4.8(a) and Fig. 4.8(b), respectively. Taking an optimistic point of view, we can say the magnitude responses show satisfactory agreement, particularly at the lower frequencies. Indeed, better agreement can be had at higher frequencies if the sampling frequency is increased, but the cost is more samples. The dotted lines in the magnitude plots of Figs. 4.8(c)-(d) show the squared-magnitude of $1/f$ with DC value set to unity. These results are consistent with the previous discussion of one-way diffraction measured by an infinite receiver but require some explanation.

Consider the beam of a finite transmitter with radius $a \gg \lambda$. In the near-field, the beam of such a transmitter may be considered a geometric extension of the transmitter face [60]. Next, consider a finite receiver of radius $b = a$ that is parallel to and centered in the beam of the transmitter. In the near field, this finite receiver will detect virtually all the transmitted energy because the beam is essentially a geometric extension of the transmitter face [5].

In other words, the receiver fits or covers the lateral extent of the beam and is indistinguishable from an infinite receiver [96]. Thus, both the arccos formulation (dashed line) and Lommel formulation (solid line) as depicted in Fig. 4.8(c) show excellent agreement with the theoretically infinite receiver case (dotted line). Farther out from the transmitter, the beam begins to spread or diffract. A finite receiver placed further away from the transmitter will no longer be indistinguishable from an infinite receiver. Thus, both the arccos formulation (dashed line) and Lommel formulation (solid line) as depicted in Fig. 4.8(d) show poor agreement with the theoretically infinite receiver case (dotted line). This insight is further testimony to the practical and theoretical value of the proposed theory of spatially averaged diffraction correction.

The phase responses do not agree as favorably. This is not surprising when one considers the physical origins of the results being compared. Specifically, the arccos-derived results are based on the Rayleigh-Sommerfeld diffraction integral, while the Lommel-derived results are based on the Fresnel diffraction integral. Hence, the two diffraction integrals differ primarily in terms of their phase [34]. This, in conjunction with Gibb's phenomenon, helps explain the phase differences exhibited in the plots. Reasons for the spectral discontinuity at the Nyquist frequency are not fully understood.

It is crucial to note, however, that Eq. 4.11 was derived under the assumption of an ideal piston transducer with a Dirac response. Thus, Eq. 4.11 is completely general in terms of frequency. Real transducers, however, are bandlimited. This observation also holds for Eq. 4.10 and Eq. 4.17 and leads to the following discussion.

Consider a real 2.25-MHz unfocused piston transducer with diameter $2a = 13$ mm. A typical bandwidth for such a transducer is 2 to 4 MHz centered at 2.25 MHz. Clearly, the frequency-domain *and* time-domain results shown in Fig. 4.8 agree quite well over that bandwidth, particularly in terms of magnitude. Thus, the closed-form frequency-domain formalism proposed here applies to the real transducer just described with just 2X oversampling. Thus, if a diffraction correction were desired for this transducer, Eq. 4.11 could be used to calculate an inverse filter directly in the frequency-domain. Furthermore, higher sampling rates could be used, and the results applied to real transducers operating at frequencies higher than 2.25 MHz.

Figs. 4.6–4.7 and Figs. 4.9–4.10 show results for cases when $b \neq a$. With the exception of discontinuities, the results show remarkable qualitative and quantitative agreement, and the discussion of results obtained for $b = a$ apply.

Indeed, the spatially averaged results for $b = a/1000$ shown in Fig. 4.6 are consistent with the results predicted by point-receiver theory [86]. In particular, the case $b = a/1000$ approximates the case of an on-axis ($\rho = 0$) point receiver, and Eq. 3.13 predicts that the velocity-potential impulse response $h_1(\rho, z, t)$ for this case should resemble a rectangular pulse. Spatially averaged impulse responses $\langle \hat{h}_1(z, t) \rangle_b$ computed with Eq. 4.10 for $b = a/1000$ are consistent with this point-receiver theory. The temporal duration of the Lommel-based impulse responses differ slightly from the arccos-based impulse responses for $b > a$ because Eq. 4.18 is based on the Fresnel approximation while Eq. 4.5 is based on the Rayleigh-Sommerfeld diffraction integral.

The results shown in Fig. 4.8 place the difference between spatial averaging and spatial integration [41, 96] in new perspective. The spatially averaged results shown in the figure are consistent in every respect with the theory of one-way diffraction for a point receiver presented in the previous chapter. See in particular Fig. 3.2 (a)–(b) and note the amplitude of the pulses. Spatially integrated results would differ by a factor of πb^2.

Finally, we note that Eq. 4.5 is superior to Eq. 4.18 if a time-domain impulse response is preferred for, say, a time-deconvolution diffraction correction [14]. Furthermore, Eq. 4.5 can certainly be used in the frequency domain given the ease with which the FFT can be performed. However, Eq. 4.18 can be calculated directly in the frequency domain across any bandwidth of interest and produces results comparable to Eq. 4.5. Thus, we have demonstrated the utility of the proposed frequency-domain formalism for spatially averaged one-way diffraction for an unfocused piston transducer. As in the time-domain case, fully coherent frequency-domain expressions for spatially averaged one-way diffraction from a focused piston transducer remain unsolved.

7.   COMPUTATIONAL CONSIDERATIONS

The computational issues discussed in Chapter 3 apply here. Analog frequency was used to compute Eq. 4.18, and the results were multiplied by Δf prior to inverse Fourier transforming. Finally, the highest-order Bessel function used in computations of X_2 and Y_2 was $J_{63}(x)$.

8.   CHAPTER SUMMARY

Closed-form time-domain and frequency-domain expressions applicable to *unfocused* one-way diffraction with a finite receiver of any radius were derived. The time-domain expressions were derived by interpreting the spatially averaged arccos diffraction formulation, in integral form, as a Hankel transform and borrowing results derived by Gaskill [34]. The time-domain results turned out to be the same as results derived by Cassereau, *et al.* [14], but the derivation and ultimate interpretation were different. On the other hand, the frequency-domain results were derived by direct spatial integration and averaging of the Lommel diffraction formulation. Detailed analysis of both the time- and frequency-domain expressions led to a number of new insights concerning diffraction from a circular aperture, and these insights have broad applicability.

Results obtained from the time- and frequency-domain expressions verified the theoretical prediction that the approximate Fourier equivalence of the arccos and Lommel diffraction formulations can be extended to spatially averaged one-way diffraction. The results also showed remarkable agreement for $b \leq a$ but began to disagree with $b > a$. This disagreement was, however, reasonably explained by the theory.

Overall, the results obtained in this chapter cause us to be optimistic that the approximate Fourier equivalence of the arccos and Lommel diffraction formulations can be extended to the autoconvolution interpretation of two-way diffraction involving a flat plate. Finally, the one-way results derived in this chapter can be applied to the mirror-image interpretation of two-way diffraction simply by doubling z in Eq. 4.5 and Eq. 4.11 when $b = a$.

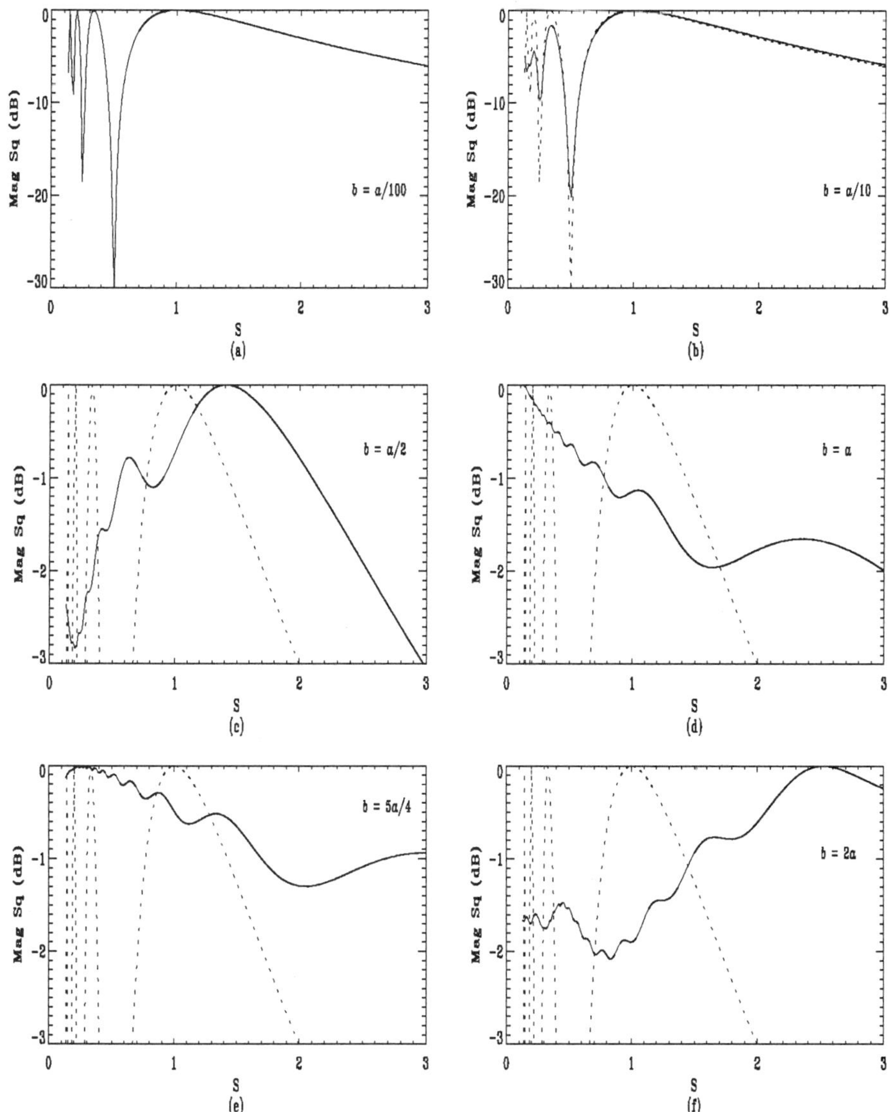

Figure 4.5.   Spatially averaged (solid lines) and on-axis (dotted lines) diffraction fields from a circular aperture of radius a measured by a receiver of radius b.

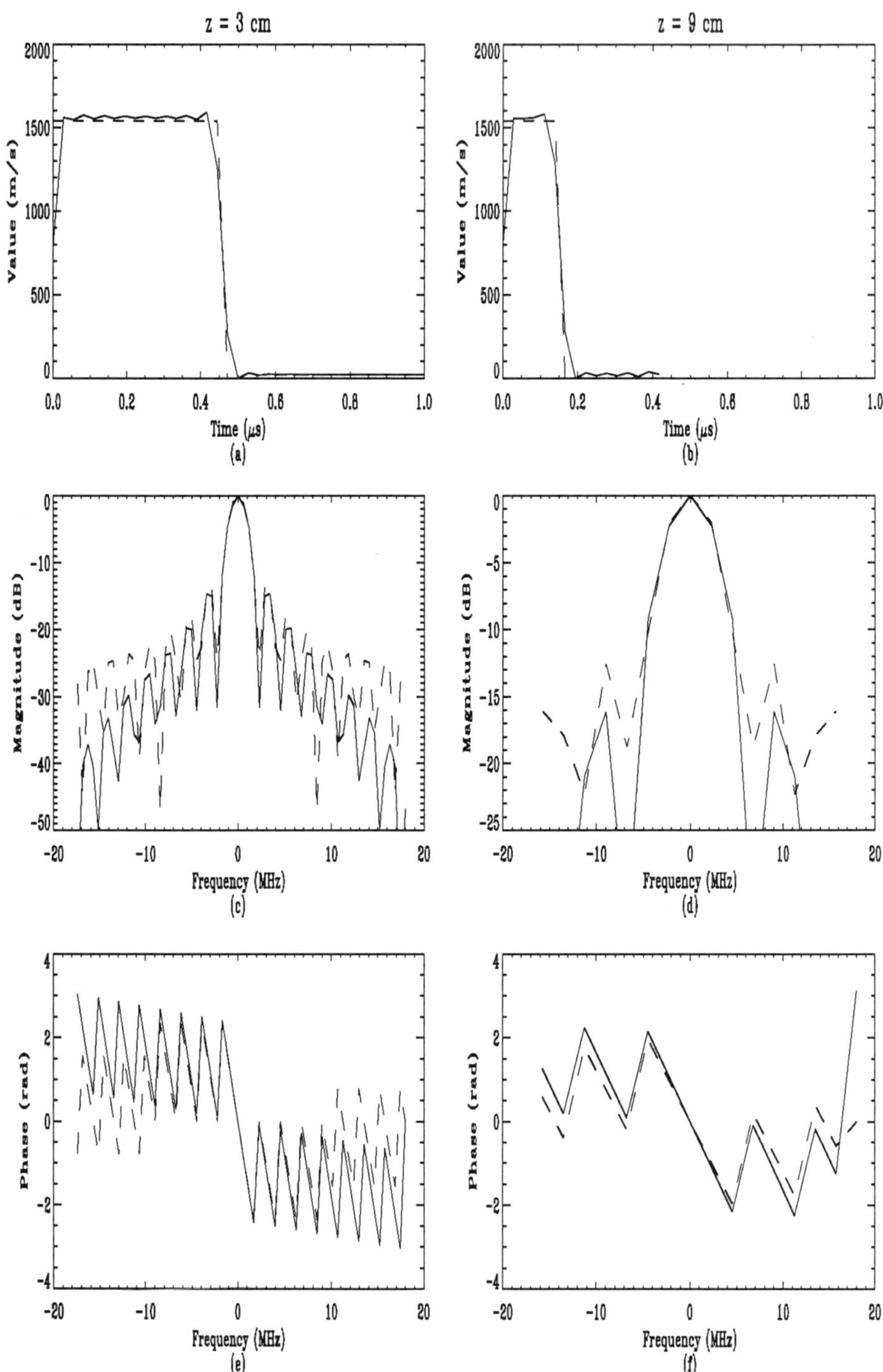

Figure 4.6. One-way spatially averaged impulse responses for the Lommel (solid) and arccos (dashed) diffraction formulations: $b = a/1000$.

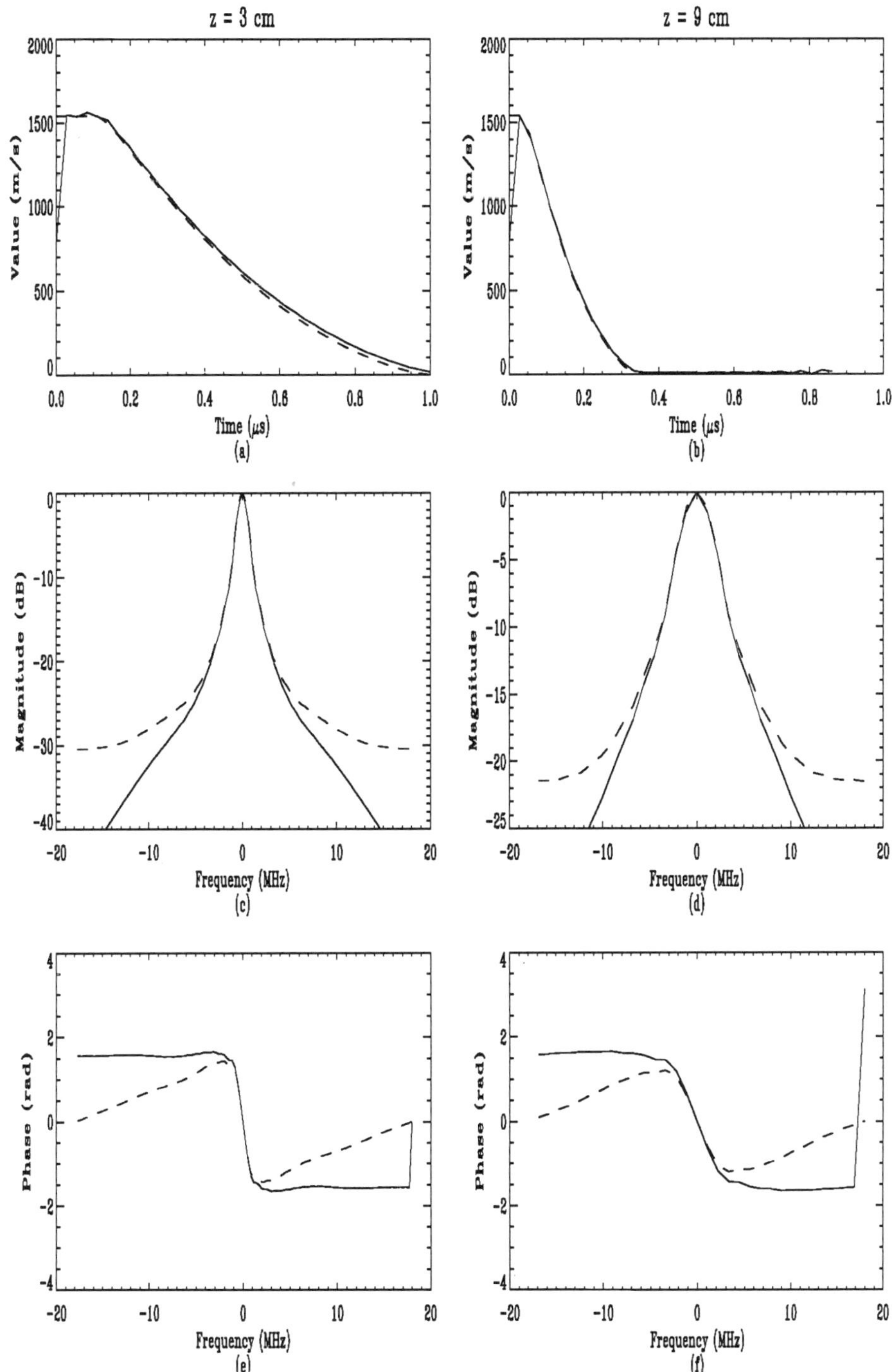

Figure 4.7. One-way spatially averaged impulse responses for the Lommel (solid) and arccos (dashed) diffraction formulations: $b = a/2$.

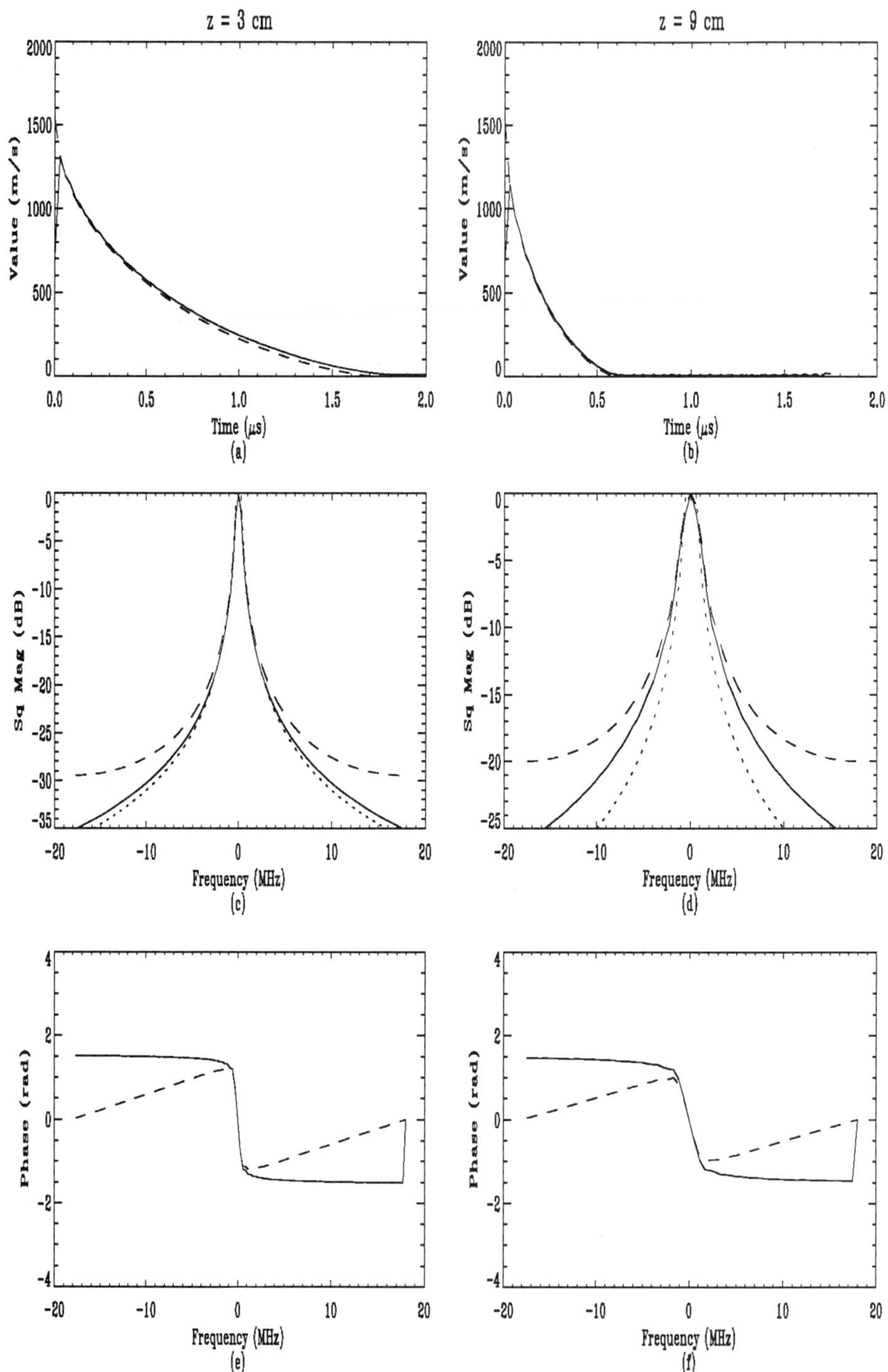

Figure 4.8.   One-way spatially averaged impulse responses for the Lommel (solid) and arccos (dashed) diffraction formulations: $b = a$.

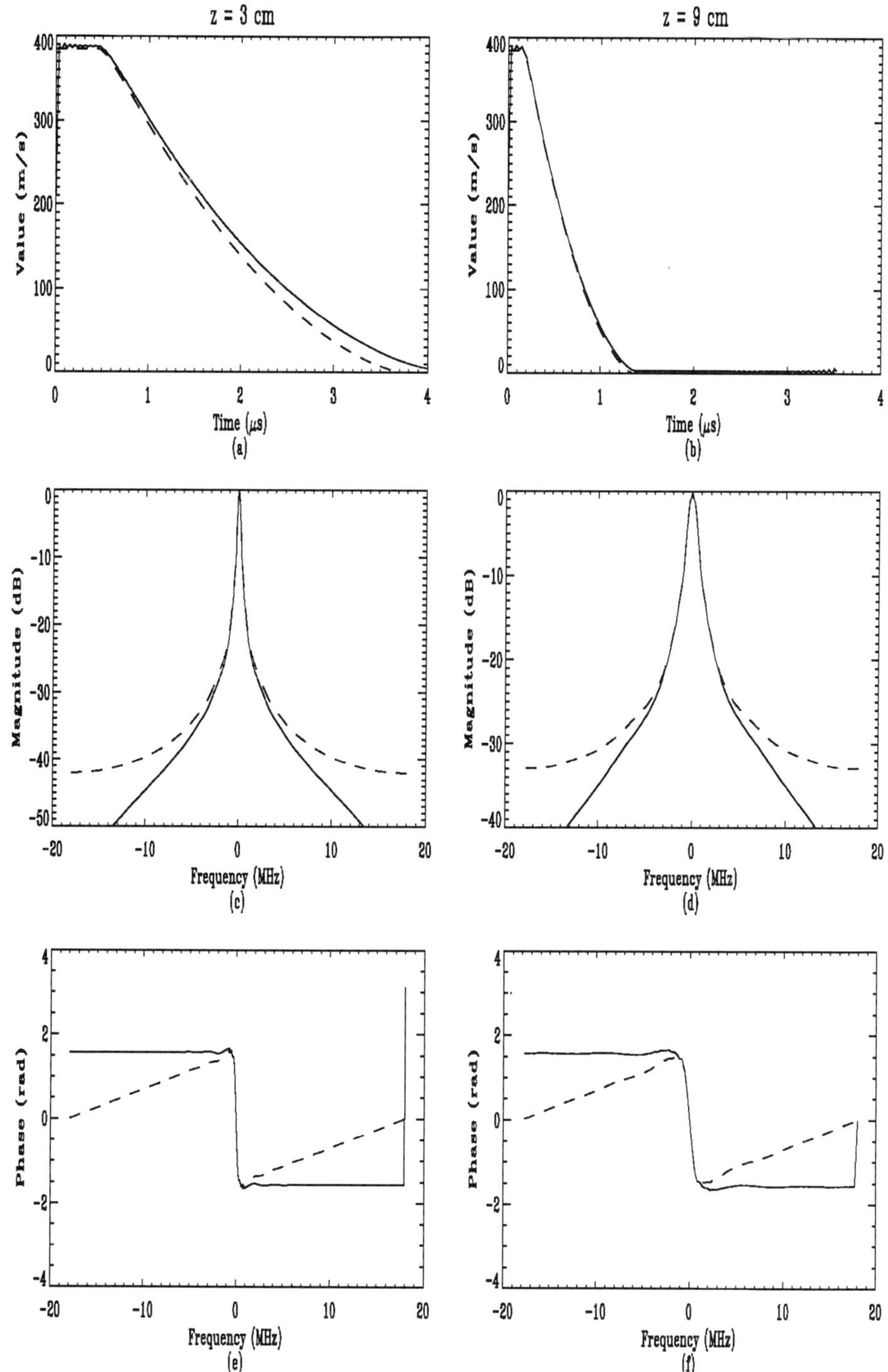

Figure 4.9. One-way spatially averaged impulse responses for the Lommel (solid) and arccos (dashed) diffraction formulations: $b = 2a$.

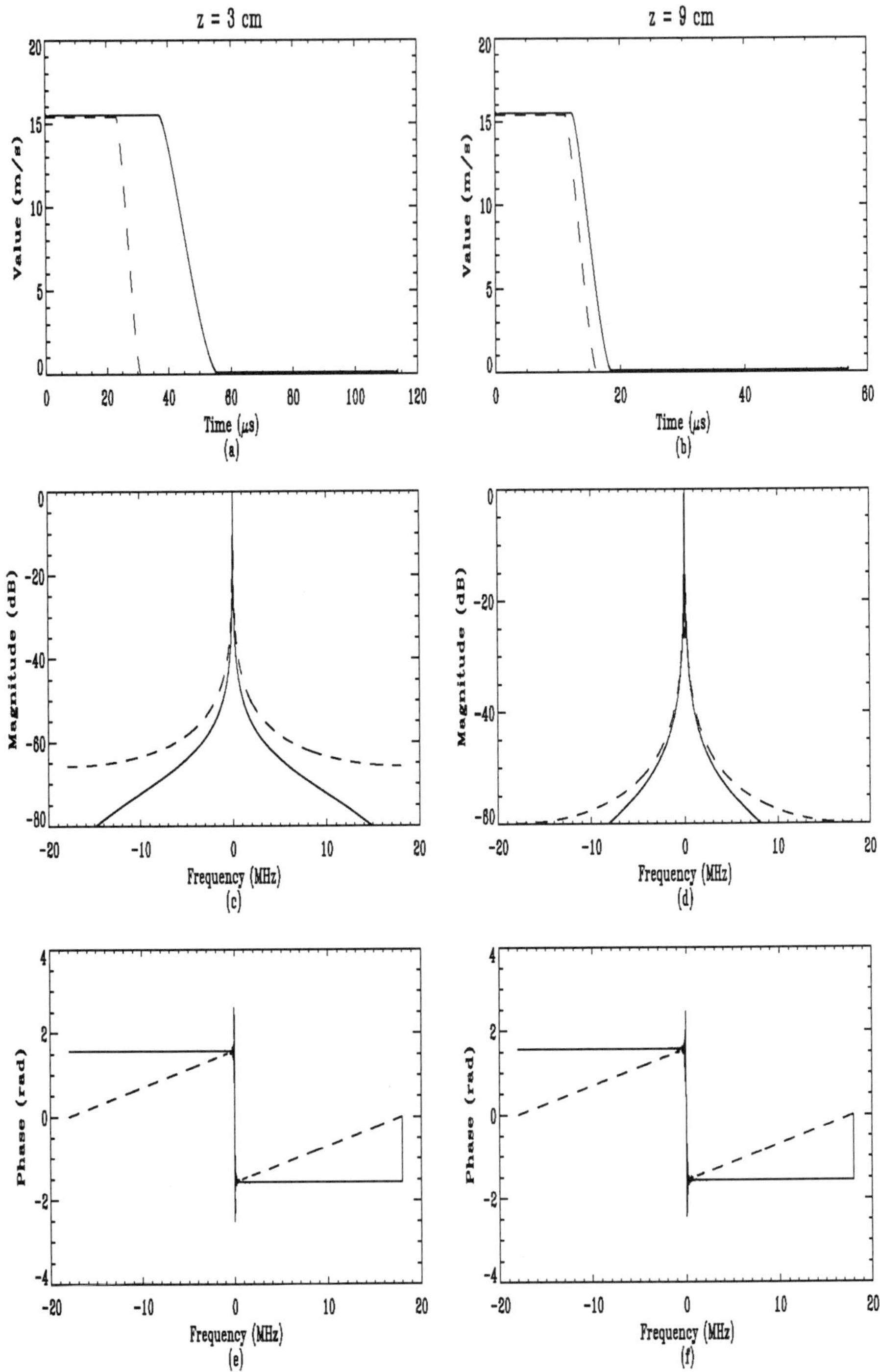

Figure 4.10. One-way spatially averaged impulse responses for the Lommel (solid) and arccos (dashed) diffraction formulations: b = 10a.

APPENDIX 4.A: Two Lemmas

Lemma 1.

$$\sum_{s=0}^{\infty}(-1)^s U_{n+2s}(u,v) = Z_n(u,v) \qquad (4.A.1)$$

Proof.

$$\sum_{s=0}^{\infty}(-1)^s U_{n+2s}(u,v) = \sum_{s=0}^{\infty}(-1)^s \sum_{p=0}^{\infty}(-1)^p \left(\frac{u}{v}\right)^{n+2s+2p} J_{n+2s+2p}(v)$$

$$(4.A.2)$$

$$= \sum_{s=0}^{\infty}\sum_{p=0}^{\infty}(-1)^{s+p} \left(\frac{u}{v}\right)^{n+2s+2p} J_{n+2s+2p}(v)$$

Collecting terms of the same order in $J_n(v)$ and arranging the series in ascending order, the lemma follows. The grouping of terms is justified since the series is absolutely convergent. $\qquad\square$

Lemma 2.

$$\frac{1}{u}\int_u^{v_b} vU_n(u,v)e^{-j\frac{v^2}{2u}}\,dv = \left\{ e^{-j\frac{u}{2}}\left[Z_{n+1}(u,u) - jZ_n(u,u)\right] \right. +$$

$$\left. e^{-j\frac{v_b^2}{2u}}\left[jZ_n(u,v_b) - Z_{n+1}(u,v_b)\right] \right\} \qquad (4.A.3)$$

Proof.

$$\frac{1}{u}\int_u^{v_b} vU_n(u,v)e^{-j\frac{v^2}{2u}}\,dv = \frac{1}{u}\int_u^{v_b} v\left\{ \sum_{s=0}^{\infty}(-1)^s \left(\frac{u}{v}\right)^{n+2s} J_{n+2s}(v) \right\} e^{-j\frac{v^2}{2u}}\,dv$$

$$= \sum_{s=0}^{\infty}(-1)^s \int_u^{v_b} \left(\frac{v}{u}\right)^{1-(n+2s)} J_{n+2s}(v)e^{-j\frac{v^2}{2u}}\,dv$$

$$(4.A.4)$$

The interchanging of the summation and integration sign above is justified since the series under the integral sign is uniformly convergent in the range of integration. A close inspection of Watson's outline of Lommel's reciprocation formulae [90, p. 543], in conjunction with Euler's theorem [100], reveals that

$$\frac{d}{dt}\left[je^{-j\frac{1}{2}wt^2} U_n\left(\frac{z^2}{w}, zt\right) - e^{-j\frac{1}{2}wt^2} U_{n+1}\left(\frac{z^2}{w}, zt\right) \right]$$

$$= z\left(\frac{wt}{z}\right)^{1-n} J_n(zt)e^{-j\frac{1}{2}wt^2}. \qquad (4.A.5)$$

Setting $t = v, w = 1/u$, and $z = 1$ yields

$$\frac{d}{dv}\left[je^{-j\frac{v^2}{2u}}U_n(u,v) - e^{-j\frac{v^2}{2u}}U_{n+1}(u,v)\right] = \left(\frac{v}{u}\right)^{1-n}J_n(v)e^{-j\frac{v^2}{2u}}.$$

$$(4.A.6)$$

The right-hand side of Eq. 4.A.6 and the integrand of Eq. 4.A.4 are in the same form. Thus, Lemma 2 follows from the fundamental theorem of integral calculus and application of Lemma 1. $\qquad\square$

Lemma 1 and Lemma 2 are simply extensions of Lemma 8 and Lemma 9, respectively, in [97].

Chapter 5

SPATIALLY AVERAGED
TWO-WAY DIFFRACTION

Recall that our ultimate goal is to derive a closed-form spatially averaged two-way diffraction correction for a focused piston transducer operating in pulsed mode. Also recall from Chapter 1 that certain methods of tissue characterization involve spectral analysis of RF data. Results produced by these methods are biased by the transmit and receive impulse responses of the transducer and by two-way diffraction effects. The bias can be removed via spectral normalization, and spectral normalization relies on a calibration spectrum which is obtained by insonifying a flat plate. This chapter explores the possibility of extending the theory of approximate Fourier equivalence to a spatially averaged version of autoconvolution diffraction from a flat plate.

Before continuing, we must remark that the model of spatially averaged two-way diffraction to be developed is a simple one. Indeed, purists will argue it is simply incorrect. We reply with four comments. First, our method has precedence in the published literature. Second, we cite this literature, but we also cite literature that suggests alternative models. Third, we note a famous statistician's observation that "all models are wrong, but some are useful [9]." Fourth, we show the potential utility of the proposed model by using it to develop new analytical insights and by applying it to real ultrasonic data.

1. SPATIALLY AVERAGED
ARCCOS DIFFRACTION FORMULATION

We claimed in Chapter 1 that it is theoretically possible to compute the required spectral normalization for any depth z with the measurement of only one calibration spectrum. The procedure is repeated here for the reader's convenience. The calibration spectrum required for spectral normalization is obtained by insonifying the front surface of a large glass plate located in water at some distance z from the transducer. Compute the diffraction effects, also known as acoustic or radiation coupling [14, 16], for the same depth and divide them out from the experimental measurement. What is left is the combined transmit/receive effect of the transducer. This effect is time invariant; hence, it can be combined with a computation of two-way diffraction effects to arrive at the complete spectral normalization for any depth z of interest.

The procedure just described requires a quick and efficient method for computing the two-way diffraction effects associated with reflection from a flat plate. Derivation of such a method begins with a linear model of reflection imaging, and the model developed by Hunt, *et al.* for a point scatterer will serve nicely [44]. Adapted to our purposes, it is

$$\langle v_R(t) \rangle_{A_t} = \frac{\varrho}{A_t} \frac{\partial^2 v_T(t)}{\partial t^2} * g(t) * h_T(\mathbf{r}, t) * s(\mathbf{r}, t) * h_R(\mathbf{r}, t) \qquad (5.1)$$

where $\mathbf{r}$ is the position vector of the point scatterer, A_t is the area of the transducer, the subscripts T and R denote transmit and receive, respectively, and all the convolutions are with respect to time. The excitation voltage is $v_T(t)$, and the voltage at the output of the receiver is $v_R(t)$. The notation $\langle \ \ \rangle_{A_t}$ indicates spatial averaging over the transducer area which is required at the outset because a finite transducer is both transmitter and receiver.

The linear time-invariant transmit and receive impulse responses of the transducer are modeled by $g(t)$ which is equal to $g_T(t) * g_R(t)$. The functions $h_T(\mathbf{r}, t)$ and $h_R(\mathbf{r}, t)$ represent the transmit and receive propagation effects, respectively. The propagation effects include diffraction as captured in the velocity-potential impulse response and may also include frequency-dependent attenuation. Lastly, $s(\mathbf{r}, t)$ is the impulse response for the point scatterer.

In practice, the received voltage $v_R(t)$ depends on reflection of ultrasonic energy from some resolution cell volume V of scatterers. Spatially integrating Eq. 5.1 over the volume of scatterers yields

$$\langle v_R(t) \rangle_{A_t, V} = \frac{\varrho}{A_t} \frac{\partial^2 v_T(t)}{\partial t^2} * g(t) * \left[\int_V h_T(\mathbf{r}, t) * s(\mathbf{r}, t) * h_R(\mathbf{r}, t) \, dV \right],$$
$$(5.2)$$

where the notation $\langle \ \ \rangle_{A_t, V}$ denotes spatial averaging over the transducer area A_t and spatial integration over the scattering volume V. Since it is understood that spatial averaging over the area of the transducer is required, the subscript A_t will be dropped. Note that $s(\mathbf{r}, t)$ in our formulation is a function of both space and time.

There is a diversity of opinion in literature on the nature of the scattering impulse response. For example, Stepanishen [87] and Hunt, *et al.* [44] formulate their scattering impulse responses as functions of time. Madsen, *et al.* formulate their scattering function $N(\mathbf{r})$ as a function of space [62]. Finally, Jensen formulates his scattering function in terms of both space and time and requires both temporal and spatial convolutions in his final model [47]. The reader is also directed to Dickinson [28], Cho, *et al.* [17], and the references these authors cite for more detailed discussions of ultrasonic scattering.

For the purposes of this monograph, we assume that the scatterers represented by $s(\mathbf{r}, t)$ are fixed in space. This assumption should be sufficient justification to write

$$s(\mathbf{r}, t) = s(\mathbf{r})\alpha(t), \tag{5.3}$$

where $\alpha(t)$ and $s(\mathbf{r})$ model the frequency-dependent attenuation and complex amplitude, respectively, of the scattering impulse response.

Writing the scattering impulse as in Eq. 5.3 is tantamount to a mathematical decoupling of space and time. Fink and Cardoso hint at this decoupling in their description of two approaches to ultrasonic propagation: the *local observer* approach and the *instantaneous* approach [31]. Given the diversity of opinion on the nature of ultrasonic scattering evident in the literature, our explicit decoupling of space and time should not be criticized too harshly. Indeed, as will be shown, it leads to equations that are consistent with the published literature.

As a result of the previous discussion, Eq. 5.2 may be written

$$\langle v_R(t) \rangle_V = \frac{\varrho}{A_t} \frac{\partial^2 v_T(t)}{\partial t^2} * g(t) * \left[\int_V h_T(\mathbf{r}, t) * s(\mathbf{r})\alpha(t) * h_R(\mathbf{r}, t)\, dV \right]. \tag{5.4}$$

The heuristic development of Eq. 5.4 is, for the most part, consistent with the more rigorous development of a pulse-echo signal model presented in an invited paper by Fink and Cardoso [31, Eq. 65].

At this point it is important to recall that our ultimate goal is to derive a closed-form spatially averaged two-way diffraction correction for a focused piston transducer operating in pulsed mode. Since diffraction corrections can be obtained by insonifying a flat plate in water [59, 62], we modify our model to account for these very common experimental conditions.

In linear acoustics, water is assumed to present no appreciable frequency dependent attenuation. Thus, the propagation effects, $h_T(\mathbf{r}, t)$

and $h_R(\mathbf{r}, t)$, in water are due solely to diffraction. This and Helmholtz's reciprocity theorem [51] allow us to write $h_T(\mathbf{r}, t) = h_R(\mathbf{r}, t) = h_1(\mathbf{r}, t)$.

A flat plate can be modeled as an acoustic impedance discontinuity with a scattering impulse response that is independent of frequency. Since the scattering impulse response is independent of frequency, $\alpha(t) = \delta(t)$ and $s(\mathbf{r}, t) = s(\mathbf{r})\delta(t)$. Thus, Eq. 5.4 becomes

$$\langle v_R(t) \rangle_V = \frac{\varrho}{A_t} \frac{\partial^2 v_T(t)}{\partial t^2} * g(t) * \left[\int_V s(\mathbf{r}) h_2(\mathbf{r}, t) \, dV \right], \qquad (5.5)$$

where the substitution

$$h_2(\mathbf{r}, t) = h_1(\mathbf{r}, t) * h_1(\mathbf{r}, t) \qquad (5.6)$$

has been used for convenience.

Note that $h_2(\mathbf{r}, t)$ models two-way diffraction effects as an autoconvolution of the one-way diffraction effects: $h_1(\mathbf{r}, t) * h_1(\mathbf{r}, t)$. Eq. 5.6 is central to the autoconvolution interpretation of two-way ultrasonic diffraction [30, Eq. 12]. It is important to note that the temporal Fourier transform of the bracketed integral in Eq. 5.5 is virtually identical to the spatial integral in a frequently cited model developed by Madsen, *et al.* [62, Eq. 3].

The discontinuous nature of the flat plate can be modeled as a deterministic and infinitely thin continuum of identical point scatterers [59, Eq. 15]:

$$s(\mathbf{r}) = \kappa \delta(\mathbf{r} - z), \qquad (5.7)$$

where κ is the reflection coefficient, which we assume is unity, and $\delta(\mathbf{r} - z)$ is a Dirac delta function. Under these assumptions, Eq. 5.5 becomes

$$\langle v_R(t) \rangle_{A_p} = \frac{\varrho}{A_p} \frac{\partial^2 v_T(t)}{\partial t^2} * g(t) * \left[\int_A h_2(x, y, z, t) \, dA \right], \qquad (5.8)$$

where z is fixed and A_p is the area or backscattering cross-section of the flat plate. Note that $\mathbf{r}$ has been rewritten in terms of x, y, and z.

Two immediate and valid criticisms may be levied against the autoconvolution interpretation of two-way ultrasonic diffraction as developed so far. First, it does not rigorously account for boundary conditions. The criticism is acknowledged, but it is also noted that boundary conditions have been simplified or neglected elsewhere in the literature [27, 48, 93]. A second and related criticism is that the autoconvolution interpretation implies the use of the *free-space* Green's function, weighted by a scattering function, to describe ultrasonic reflection as a simple linear superposition of two-way diffraction effects by point-like scatterers. Again,

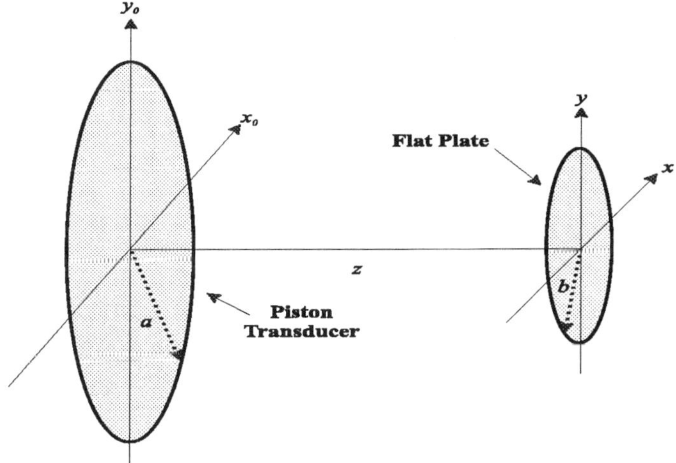

Figure 5.1.   Piston transducer and reflecting plate.

the criticism is acknowledged, but it is also noted that many authors have relied on this interpretation, either implicitly or explicitly, for over a decade [15, 27, 32, 62].

Alternative developments relying on precise accounting of boundary conditions may be found in papers by Bozma and Kuc [10], Jensen [47], and Lizzi, *et al.* [59]. It is important to note, however, that even though they rigorously accounted for boundary conditions, Lizzi, *et al.* ultimately ended up with a spatial integral [59, Eq. 16] that is virtually identical to the Fourier transform of the spatial integral in Eq. 5.8.

We now focus on the bracketed integral in Eq. 5.8 and limit our attention to a piston transducer with radius a and a circular reflecting plate with radius b. The situation is illustrated in Fig. 5.1. Circular symmetry allows us to write the integral in question as

$$\langle h_2(z,t)\rangle_b = \frac{1}{\pi a^2}\left[2\pi \int_0^b h_2(\rho,z,t)\,\rho\,d\rho,\right] \tag{5.9}$$

where spatial averaging by the transducer area $A_t = \pi a^2$ has been incorporated and the transformation to cylindrical coordinates is obvious. The notation $\langle\ \rangle_b$ indicates spatial integration over the face of an infinitely thin reflecting plate with radius b. Note the angular integration from 0 to 2π has been completed.

At this point it is important to note that the calibration spectrum discussed at the beginning of the chapter is based on reflection from a *infinitely* large reflecting plate. Strictly speaking then, the upper limit in the integral of Eq. 5.9 should be infinity. For the sake of generality, we leave the upper limit *finite* and unspecified. This decision will prove beneficial.

Expanding the integrand in Eq. 5.9 in terms of the arccos diffraction formulation $h_1(\rho, z, t)$ yields

$$\langle h_2(z,t)\rangle_b = \frac{1}{\pi a^2}\left[2\pi\int_0^b h_1(\rho,z,t) * h_1(\rho,z,t)\ \rho\ d\rho\right]. \tag{5.10}$$

Thus, Eq. 5.10 may be interpreted as an autoconvolution form of the acoustic or radiation coupling described in Chapter 1 and elsewhere [16, 76]. It would be ideal to have a closed form solution to this integral and then use it to compute complete spectral normalizations for the piston transducer being used. Unfortunately, Eq. 5.10 is simply too complicated to solve in closed form.

2.   SPATIALLY AVERAGED LOMMEL DIFFRACTION FORMULATION

Our initial attempt at deriving a spatially averaged form of autoconvolution diffraction was thwarted by the extraordinarily difficult time-domain integral in Eq. 5.10. A counter-attack in the frequency domain will lead to a limited but nonetheless satisfactory and respectable victory. In this section, it will be shown that magnitude-only expressions based on Wolf's treatment of monochromatic optical diffraction [97] can be can be adapted to ultrasonic piston transducers operating in pulsed mode. The assumption of minimum phase will enable approximation of a phase response which, in turn, will allow estimation of the spatially averaged two-way impulse for autoconvolution diffraction.

The temporal Fourier transform of Eq. 5.10 is

$$\langle H_2(z,\omega)\rangle_b = \frac{1}{\pi a^2}\left[2\pi\int_0^b H_1(\rho,z,\omega)\,H_1(\rho,z,\omega)\,\rho\,d\rho\right]. \tag{5.11}$$

The approximate Fourier equivalence of the arccos and Lommel formulations developed in Chapter 3 allows us to substitute the *focused* Lommel diffraction formulation $\widehat{H}_1(\rho, z, \omega)$ (Eq. 3.22) for $H_1(\rho, z, \omega)$.

We implied, in the previous section, that the calibration spectrum derived by Lizzi and co-workers was based on or, at least, closely related to the autoconvolution interpretation of ultrasonic reflection from a flat plate. Here, we explicitly state that the calibration spectrum derived by Lizzi is ultimately an autoconvolution interpretation of two-way diffraction. Specifically, the integral in Eq. 16 of Reference [59] is identical, in interpretation, to the bracketed integral in Eq. 5.11.

We must, however, point out four differences between the Lizzi result and Eq. 5.11. First, the former is written in rectangular coordinates while the latter is written in cylindrical. Second, the former is, strictly speaking, an improper integral while the latter can be made improper by letting b approach infinity. We note that the parameter b in Eq. 5.11 is a degree of freedom that will allow us to introduce an effective backscattering cross-section in the next chapter. Third, the former was derived with rigorous accounting of boundary conditions while the latter was not.

The last difference requires more detailed explanation. Eq. 16 in Reference [59] is strictly valid near the focus of a focused transducer and in the far field of an unfocused transducer. Although Lizzi, *et al.* did not so state, their Eq. 16 holds in the far field of an unfocused transducer because it is based on O'Neil's results which do [67]. We also note that Eq. 5.11 is essentially a slice of the coherent scattering term D developed by Madsen, *et al.* for a piston transducer [62]. Thus, the autoconvolution diffraction result in Eq. 5.11 appears to be fairly well-established in the literature.

The integral that results from substituting $H_1(\rho, z, \omega)$ with $\widehat{H}_1(\rho, z, \omega)$ in Eq. 5.11 is still complicated because of the implicit demand to retain the phase of the two-way transfer function. Recall that phase is required to compute an impulse response via an inverse Fourier transform. If the demand for phase information is dropped, we may be able to obtain some meaningful closed-form result. Dropping the demand for phase information allows us to explore the squared magnitude of Eq. 5.11:

$$\left| \langle \widehat{H}_2(z, \omega) \rangle_b \right|^2 = \left| \frac{1}{\pi a^2} \left[2\pi \int_0^b \widehat{H}_1(\rho, z, \omega) \, \widehat{H}_1(\rho, z, \omega) \, \rho \, d\rho \right] \right|^2. \tag{5.12}$$

The Cauchy-Schwarz inequality for integrals tells us that

$$\left| \frac{1}{\pi a^2} \left[\int_0^b \widehat{H}_1^2(\rho, z, \omega) \, \rho \, d\rho \right] \right|^2 \leq \left(\frac{1}{\pi a^2} \left[2\pi \int_0^b \left| \widehat{H}_1(\rho, z, \omega) \right|^2 \rho \, d\rho \right] \right)^2.$$
$$\tag{5.13}$$

The Cauchy-Schwarz inequality also indicates that equality in Eq. 5.13 *does not* hold. This discussion and Eqs. 5.12–5.13 allow us to write

$$\left| \langle \widehat{H}_2(z, \omega) \rangle_b \right|^2 < \left(\frac{1}{\pi a^2} \left[2\pi \int_0^b \left| \widehat{H}_1(\rho, z, \omega) \right|^2 \rho \, d\rho \right] \right)^2. \tag{5.14}$$

Taking the square root of Eq. 5.14 yields

$$\left| \langle \widehat{H}_2(z, \omega) \rangle_b \right| < \frac{1}{\pi a^2} \left[2\pi \int_0^b \left| \widehat{H}_1(\rho, z, \omega) \right|^2 \rho \, d\rho \right]. \tag{5.15}$$

Eq. 5.15 provides an upper bound for the magnitude response associated with the spatially averaged transfer function for autoconvolution diffraction involving a flat plate.

This upper bound can be interpreted mathematically by letting

$$f(\rho, z, t) = h_1(\rho, z, t) \star h_1(\rho, z, t), \qquad (5.16)$$

where $\star$ denotes correlation over time t. Thus, $f(\rho, z.t)$ is the autocorrelation of the one-way diffraction impulse response $h_1(\rho, z, t)$. Spatially averaging $f(\rho, z, t)$ and taking its Fourier transform yields

$$\mathcal{F}\left\{ \frac{1}{\pi a^2} \left[2\pi \int_0^b f(\rho, z, t)\, \rho\, d\rho \right] \right\} \approx \frac{1}{\pi a^2} \left[2\pi \int_0^b \left| \widehat{H}_1(\rho, z, \omega) \right|^2 \rho\, d\rho \right],$$
$$(5.17)$$

where the approximate Fourier equivalence of the arccos and Lommel diffraction formulations has been invoked. Note that the right-hand sides of Eq. 5.15 and Eq. 5.17 are equal. Thus, the frequency-domain magnitude of the spatially averaged, one-way velocity-potential *autoconvolution* is strictly less than the frequency-domain magnitude of the spatially averaged, one-way velocity-potential *autocorrelation*.

With this insight in mind, we make the *ad hoc* assumption that the frequency-domain magnitude of the spatially averaged autoconvolution is approximately equal to the frequency-domain magnitude of the spatially averaged autocorrelation:

$$\left| \langle \widehat{H}_2(z, \omega) \rangle_b \right| \approx \frac{1}{\pi a^2} \left[2\pi \int_0^b \left| \widehat{H}_1(\rho, z, \omega) \right|^2 \rho\, d\rho \right]. \qquad (5.18)$$

Our results and a discussion of coherent and incoherent averaging in Chapter 7 will help justify the assumption. The literature provides further justification. Specifically, the right-hand side Eq. 5.18 is essentially Eq. 74 in Reference [31] with the finite window and Fourier transform of the arccos diffraction formulation replaced by a Dirac delta function and the Lommel diffraction formulation, respectively.

With $u = ka^2/|\epsilon|$, $v = ka\rho/z$, and $v_b = kab/z$, Eq. 5.18 becomes

$$\left| \langle \widehat{H}_2(z, \omega) \rangle_b \right| \approx \frac{1}{k^2} \left\{ \frac{2}{u^2} \int_0^{v_b} [U_1^2(u, v) + U_2^2(u, v)]\, v\, dv \right\}. \qquad (5.19)$$

Wolf solved the braced integral in Eq. 5.19 for $v_b < u$, $v_b = u$, and $v_b > u$ [97]. Wolf's solution is long and complicated, and the reader is referred to the original reference for details. The results for the three regions are for $v_b < u$,

$$\left| \langle \widehat{H}_2(z,\omega) \rangle_b \right| \approx \frac{1}{k^2} \left(\left(\frac{v_b}{u} \right)^2 \left[1 + \sum_{s=0}^{\infty} \frac{(-1)^s}{2s+1} \left(\frac{v_b}{u} \right)^{2s} Q_{2s}(v_b) \right] \right.$$

$$\left. - \frac{4}{u} \left[Y_1(u, v_b) \cos\left(\frac{u}{2} + \frac{v_b}{2u} \right) + Y_2(u, v_b) \sin\left(\frac{u}{2} + \frac{v_b}{2u} \right) \right] \right), \quad (5.20)$$

for $v_b = u$,

$$\langle \widehat{H}_2(z,\omega) \rangle_b \approx \frac{1}{k^2} \left(1 - J_0(u) \cos(u) - J_1(u) \sin(u) \right), \quad (5.21)$$

and for $v_b > u$,

$$\langle \widehat{H}_2(z,\omega) \rangle_b \approx \frac{1}{k^2} \left(1 - \sum_{s=0}^{\infty} \frac{(-1)^s}{2s+1} \left(\frac{u}{v_b} \right)^{2s} Q_{2s}(v_b) \right). \quad (5.22)$$

These results reduce to the case of an unfocused transducer in the limit as $A \to \infty$ since $\lim_{A \to \infty} \epsilon = z$. Recall A is the focal length.

3.   ANALYSIS OF FREQUENCY-DOMAIN RESULTS

In this section, the frequency-domain expressions in Eqs. 5.20–5.22 are examined. We are particularly interested in their behavior as a function of b, the radius of the reflecting disk.

Figs. 5.6–5.17 show attenuation due to diffraction (the diffraction filter) as a function of frequency f and depth z for different values of the parameter b and for different types of focusing. Relevant parameters are annotated in the figures. The figures are grouped at the end of the chapter for convenience. Recall from Chapter 3 that short (strong) focusing implies $A \approx 0.2a^2/\lambda$, medium focusing implies $A \approx 0.5a^2/\lambda$, and long (weak) focusing implies $A \approx 0.8a^2/\lambda$ where a is the radius of the transducer [52, 69].

As before, the speed of sound was set at $c = 1540$ m/s, and piston diameters were set at $2a = 13$ mm. The transducer was assumed to have an infinitely broadband response, and the excitation was assumed to be an impulse. The sampling frequency was set at $f_S = 36$ MHz; thus, the Nyquist frequency was 18 MHz. The annotation $f = 2.25$ in the plots is a reminder that sampling rate was set based on 2X oversampling of real piston transducer with $f_c = 2.25$ MHz and an upper frequency of 4.5 MHz.

The figures show how attenuation due to two-way diffraction varies as a function of depth, frequency, focusing, and reflecting plate radius.

First, we note general trends and then move onto more specific observations. The figures confirm the two well-known facts that (i) attenuation due to monochromatic diffraction increases with depth and (ii) lower frequency energy attenuates more with distance than does higher frequency energy. Note also that relative attenuation increases with increasing frequency at a fixed depth z.

Before discussing how focusing and reflecting area or backscattering cross-section figure in diffraction, we consider conservation of energy at a single frequency in terms of reflection from a plate with some finite radius b. The plate is free to move along the z-axis. For a plate of fixed radius b, intuition and theory predict that the reflected energy received by an *unfocused* transducer will oscillate in the near field, be maximized near the last axial maximum, and monotonically decrease as the plate is moved farther and farther out into the far field. As the radius of the plate is increased, we expect that more of the transmitted energy will be reflected at all depths. As a result, we expect that attenuation due to diffraction will decrease with increasing plate radius. Fig. 5.6, Fig. 5.10, and Fig. 5.14 confirm our expectations on both counts. Indeed, these results are consistent with the behavior of the unfocused echographic diffraction filter and the mean diffraction filter illustrated in Fig. 1 of Reference [13] and Fig. 14 of Reference [31], respectively.

For a plate of fixed radius b, intuition and theory predict that the reflected energy received by a *focused* transducer will be maximized when the plate is at the focus and will decrease as the plate is moved away from the focus. Thus, attenuation due to diffraction will be minimized near the focus of the transducer. Furthermore, attenuation due to diffraction at a fixed depth z should decrease as the radius b of the reflecting plate is increased because the plate will reflect more of the transmitted energy. Figs. 5.7–5.9, Figs. 5.11–5.13, and Figs. 5.15–5.17 confirm our expectations on both counts. Indeed, these results are consistent with the behavior of the focused echographic diffraction filter illustrated in Fig. 3 of Reference [13].

The focused and unfocused results shown in the figures lead to the following conclusion. With regard to reflection from a flat plate, spatial averaging smoothes out oscillations in the near field, decreases the rate of monotonic fall-off in the far field, and eventually averages the near-field out as b is increased. This effect is similar to the effect spatial averaging has in the one-way case.

The previous results were obtained by assuming a plate with some finite radius. A theoretical result and new physical insight can be gained by considering Eq. 5.22 and Fig. 5.1 in terms of an infinite plate and the total, as opposed to spatially averaged, energy it reflects back to the transducer. Multiplying Eq. 5.22 by πa^2 produces

$$(\pi a^2)\,\langle \widehat{H}_2(z,\omega)\rangle_b \approx \left(\frac{\pi a^2}{k^2} - \frac{\pi a^2}{k^2} \sum_{s=0}^{\infty} \frac{(-1)^s}{2s+1} \left(\frac{u}{v_b}\right)^{2s} Q_{2s}(v_b) \right). \quad (5.23)$$

Letting b approach infinity yields

$$\lim_{b \to \infty} (\pi a^2)\langle \widehat{H}_2(z,\omega)\rangle_b = \frac{\pi a^2}{k^2}. \quad (5.24)$$

Even thought it contains no phase information, Eq. 5.24 is in many ways similar to Eq. 4.23. The most prominent similarity is that both Eq. 4.23 and Eq. 5.24 were derived for infinite quantities. Specifically, Eq. 4.23 was derived for an infinite receiver, while Eq. 5.24 was derived for reflection from an infinite plate. Hence, our discussion of Eq. 5.24 is patterned after the discussion of Eq. 4.23 in Chapter 4.

We first note that Eq. 5.24 is proportional to the *magnitude* of the total maximum pressure reflected by an infinite plate. In other words, it is proportional to the magnitude of the total pressure reflected by an infinite disk which was disturbed by acoustic energy produced by a section of area πa^2 cut out of a plane wave that has the same particle velocity, in our case unity, as does the piston source. This logic is simply an extension of Williams' reasoning discussed in the previous chapter.

In essence, the magnitude of the total pressure detected by the transducer is the same from all z-planes. No pressure or energy is lost because (i) the reflecting disk is infinite, (ii) no loss mechanism has been introduced into the theory, and (iii) goniemetric subtleties of scattering have not been accounted for. Diffraction, in this case, probably introduces only a depth-dependent phase shift. Indeed, a close inspection of the work done by Lizzi, *et al.* [59, Eqs. 15-18] lends credibility to these assertions.

More insight can be gained by examining Eqs. 5.23–5.24. Note the first term in Eq. 5.23 is identical to right-hand side of Eq. 5.24. Thus, the first term in Eq. 5.23 represents the infinite-reflector solution, and the second term represents the influence of diffraction [5]. Similar observations can be made about Eqs. 5.20–5.21. Furthermore, the $1/k^2$ factor in the infinite-reflector solution indicates that the autoconvolution interpretation of two-way diffraction is dominated by a $1/f^2$ lowpass filtering effect. This observation is analogous to the $1/f$ low-pass filtering we saw in the one-way case.

Finally, the $1/f^2$ low-pass filtering effect for two-way diffraction is intuitively appealing: one-way diffraction squared. The result is appealing because of its symmetry, and this tempts us to claim that the depth-dependent phase shift associated with Eq. 5.24 is e^{-jk2z}. Indeed, the phase term in Eq. 17 of Reference [59] lends credibility to this claim.

4.   EXTENDING FOURIER EQUIVALENCE

We should be able to apply the theory of approximate Fourier equivalence to spatially averaged autoconvolution diffraction involving a flat plate. However, Eqs. 5.20–5.22 contain no phase information; thus, the inverse Fourier transform will be insufficient when it comes to computing a spatially averaged autoconvolution impulse response. The concept of minimum phase [74] will help overcome this problem.

In simplistic terms, a minimum-phase system is one that is causal and stable; see Rabiner and Schafer [74] for details and definitions. A general property of a minimum-phase system is that its phase response can be calculated from its magnitude response and vice versa. Since the autoconvolution of $h_1(\rho, z, t)$ is both causal and stable, a minimum-phase solution is assumed. The minimum-phase assumption is not without precedent in ultrasound. See, for example, Kuc's paper on modeling acoustic attenuation [54]. Thus,

$$\langle \hat{h}_2(z,t) \rangle_b = \mathcal{F}^{-1} \left\{ \left| \langle \widehat{H}_2(z,\omega) \rangle_b \right| e^{\pm j\phi(\omega)} \right\} \tag{5.25}$$

where the minimum phase $\phi(\omega)$ is estimated by taking the Hilbert transform of $\ln(|\langle \widehat{H}_2(z,\omega) \rangle_b|)$. The choice of sign in Eq. 5.25 depends on the depth z and accounts for the time reversal of the focused arccos diffraction formulation when $z > A$ [2]. No such time reversal occurs for the unfocused case, and the sign of the phase is chosen based on the sign convention of the FFT being used. Recall Eq. 5.25 is based on Eq. 3.22 which, as explained in Chapter 3, holds for focused and unfocused transducers; thus, Eq. 5.25 should hold, theoretically at least, for both focused and unfocused piston transducers.

5.   VERIFICATION

Results obtained from Eq. 5.25 were plotted against results computed via numerical integration of the arccos diffraction formulation. Figs. 5.2–5.5 illustrate the results. All results were normalized to a maximum value of unity for reasons which will be discussed later.

Before discussing the results, we should note well that Eq. 5.25 is an assumption in an approximation wrapped in estimation. Specifically, we have made the *ad hoc* assumption that the magnitude of $\langle \widehat{H}_2(z,\omega) \rangle_b$ is approximately equal to its upper bound as calculated by the Cauchy-Schwarz inequality. Additionally, $\langle \widehat{H}_2(z,\omega) \rangle_b$ is based on the Fresnel approximation. Finally, the phase response of $\langle \hat{h}_2(z,\omega) \rangle_b$ is estimated using the minimum-phase assumption, and the DC value of $\langle \hat{h}_2(z,t)\rangle_b$ is estimated using the method described in Chapter 3. Thus, if the normalized results shown in the figures are in reasonable agreement, we should declare victory and not necessarily quibble about how the battle was fought.

The data plotted in the figures were computed for an *unfocused* piston transmitter with diameter $2a = 13$ mm and a reflecting disk with radius b. Four different values of b were used, and these are noted in the figures. As in the one-way case, the impulse responses were calculated for two depths: $z = 3$ cm and $z = 9$ cm. The speed of sound was set at $c = 1540$ m/s, and the transducer was assumed to have an infinitely broadband response. The excitation was assumed to be an impulse. The sampling frequency was set at $f_S = 36$ MHz; thus, the Nyquist frequency was 18 MHz. Only Fig. 5.4 is discussed in detail. Concise comments pertaining to the Figs. 5.2–5.3 and Fig. 5.5 follow the discussion of Fig 5.4.

Figs. 5.4(a)–(b) show two-way spatially averaged impulse responses estimated via Eq. 5.25 (solid lines) and spatially averaged two-way impulse responses calculated by numerical integration of Eq. 5.10 (dashed lines). The results computed using Eq. 5.25 differ only slightly from the results computed by numerically integrating the autoconvolution of the arccos formulation, but otherwise the Lommel-based results capture the salient features computed by numerical integration, particularly time-compression with increasing depth. These results justify the assumptions discussed at the beginning of this section.

Figs. 5.4(c)–(d) show the squared magnitude responses (dB) associated with the impulse responses in Fig. 5.4(a) and Fig. 5.4(b), respectively. The magnitude responses show excellent agreement over a wide range of frequencies. As in the one-way case, better agreement can be had at higher frequencies if the sampling frequency is increased, but again the cost is more samples. The ease of computing the magnitude response with Eq. 5.21 must be emphasized; numerical integration and do-loops are not required. Furthermore, computation can be done directly in the frequency domain.

The claim that autoconvolution diffraction with an infinite reflector is dominated by a $1/f^2$ filtering effect is consistent with the dotted lines shown in Figs. 5.4(c)-(d), but the figures require some explanation. Consider a plate with radius $b = a$ in terms of the beam pattern of an unfocused transducer with radius a. Close in to the transducer, the beam is concentrated in a region bounded by the dimensions of the transducer. Thus, a finite reflector with radius $b = a$, as in Fig. 5.4(c), placed close to the transducer reflects most of the transmitted energy because it has the same dimensions as the transmitter. In effect, it is indistinguishable from an infinite reflector, and the dotted line in Fig. 5.4(c) is in excellent agreement with the theory. Farther out from the transducer, the beam begins to spread or diffract, and a finite reflector placed further away from the transducer will not have the same effect as an infinite reflector. Thus, the dotted line in Fig. 5.4(c) is not consistent with the theory.

Figs. 5.4(e)–(f) show the phase responses associated with the impulse responses in Fig. 5.4(a) and Fig. 5.4(b), respectively. The results show satisfactory agreement, but the computation of each phase response required a Hilbert transform. However, the Hilbert transform is a is a routine computation in signal processing which can be implemented fairly easily [101]. Finally, as in the one-way case, it is crucial to note Eqs. 5.20–5.22 were derived under the assumption of an ideal piston transducer with a Dirac response. As a result, Eqs. 5.20-5.25 are completely general in terms of frequency. But real transducers are bandlimited.

As in the one-way case, let's consider a real 2.25-MHz unfocused piston transducer with diameter $2a = 13$ mm. A typical bandwidth for such a transducer is 2 to 4 MHz centered at 2.25 MHz. Clearly, the results shown Fig. 5.4 apply to this real transducer. Indeed, they apply quite well, particularly in a magnitude sense, with only 2X oversampling. Thus, if a spatially averaged autoconvolution diffraction correction were desired for this transducer, Eq. 5.21 could be used to calculate an inverse filter directly in the frequency domain. Furthermore, higher sampling rates could be used, and the results applied to real transducers operating at higher frequencies than 2.25 MHz. Thus, we have again demonstrated the utility of the proposed theory for spatially averaged diffraction correction.

Figs. 5.2–5.3 and Fig. 5.5 show results for $v_b \neq u$. Overall, the normalized results show quite satisfactory results given the stated approximations. Indeed, the spatially averaged results for $b = a/1000$ in Fig. 5.2 are consistent with the results predicted by point-receiver theory [87]. In particular, the case $b = a/1000$ approximates the case of an on-axis ($\rho = 0$) point scatterer, and Eq. 5.6 predicts that $h_2(\rho, z, t)$ for this case should be triangular in shape as a result of convolving two rectangular pulses. Spatially averaged impulse responses $\langle \hat{h}_z(z, t) \rangle_b$ computed with Eq. 5.20 for $b = a/1000$ are consistent with the point-receiver theory.

Agreement between the two sets of unfocused results begins to break down with $b > a$ (Fig. 5.5). The behavior of Lommel-based results for $b > a$ has been noted in Chapters 3–4 and is not surprising since the Lommel diffraction formulation is based on the Fresnel approximation. Thus, we did not bother computing results for any $b > 2a$.

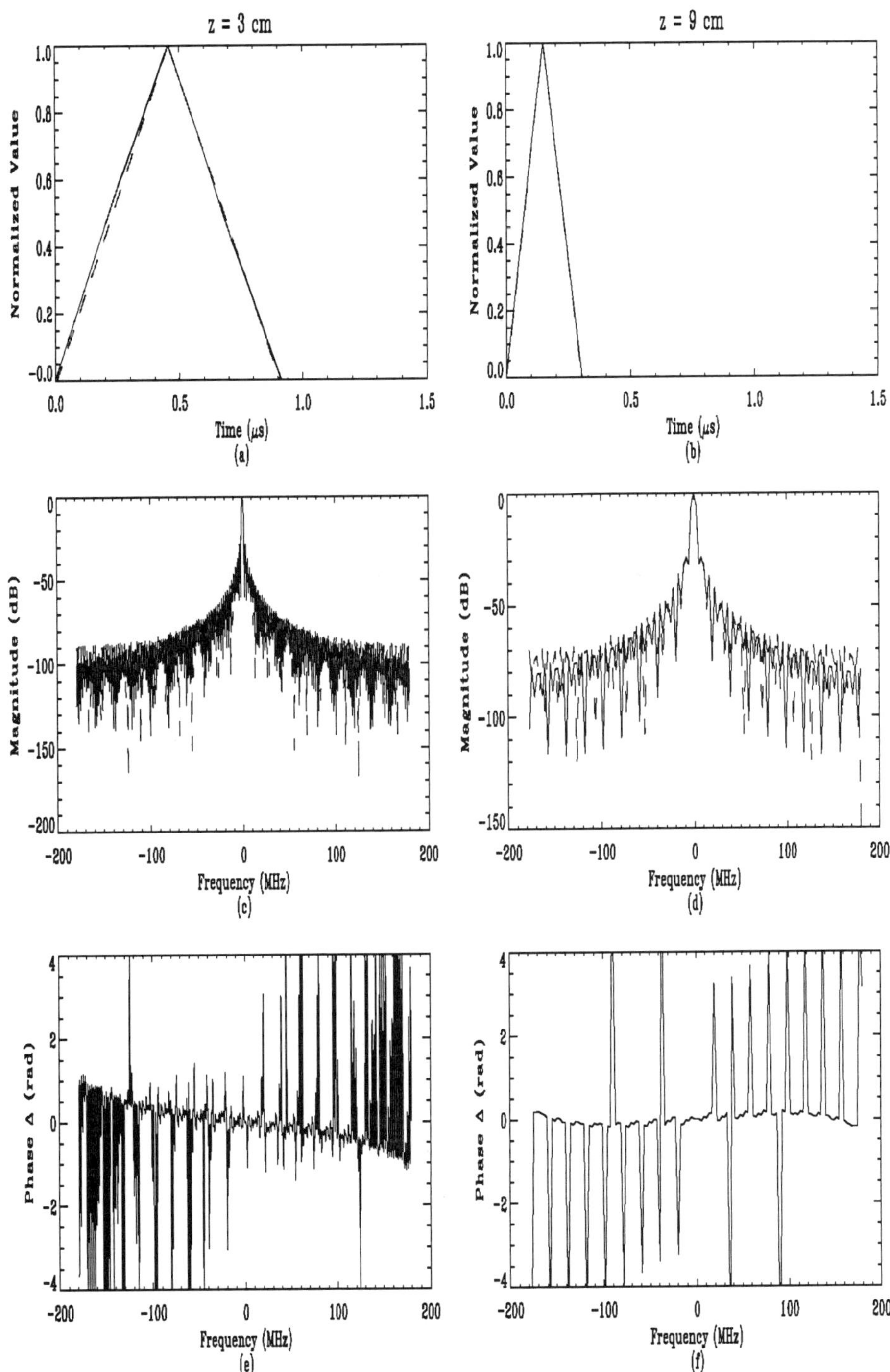

Figure 5.2. Two-way spatially averaged impulse responses for the Lommel (solid) and arccos (dashed) diffraction formulations: $b < a/1000$.

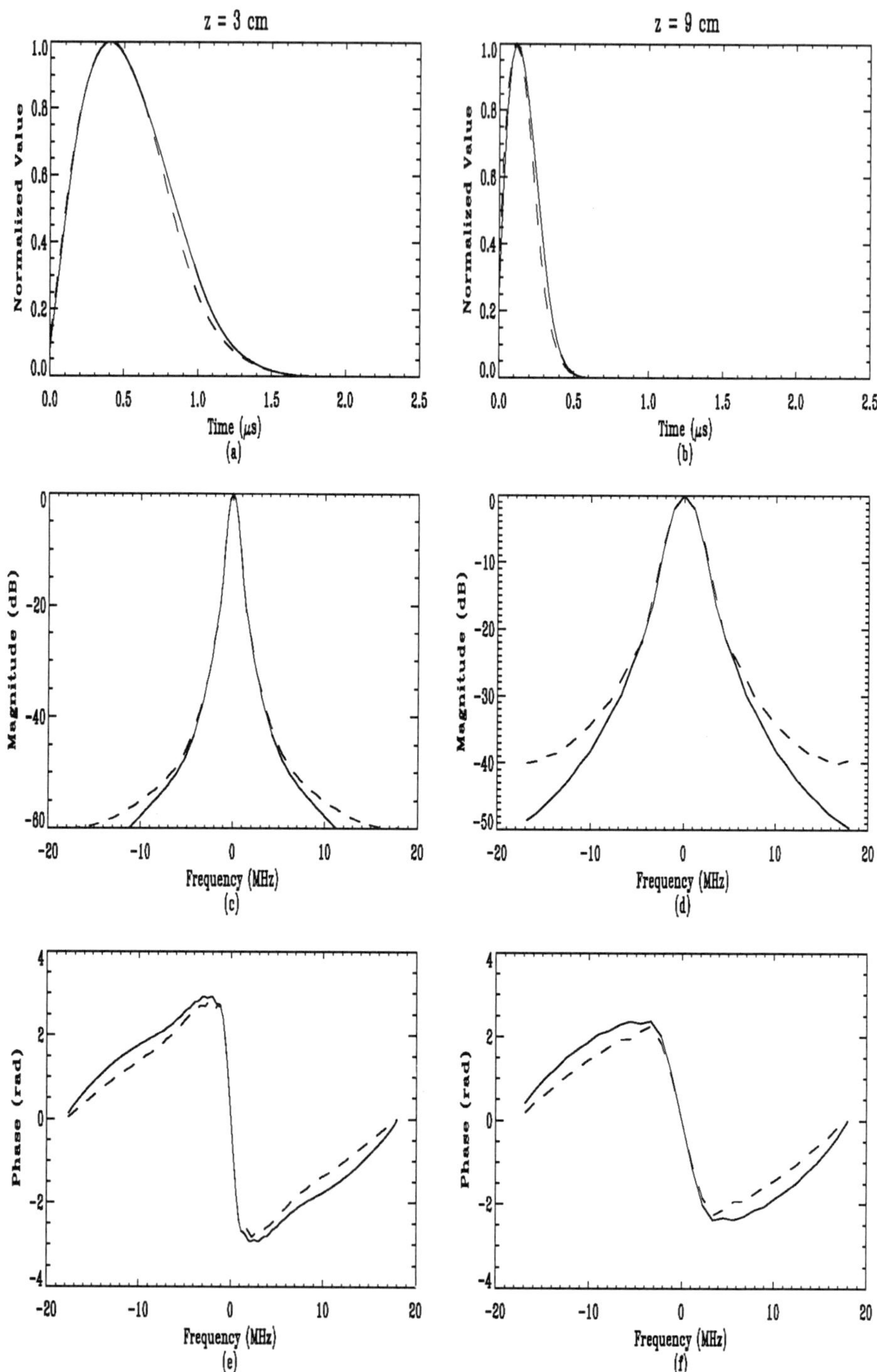

Figure 5.3. Two-way spatially averaged responses for the Lommel (solid) and arccos (dashed) diffraction formulations: $b = a/2$.

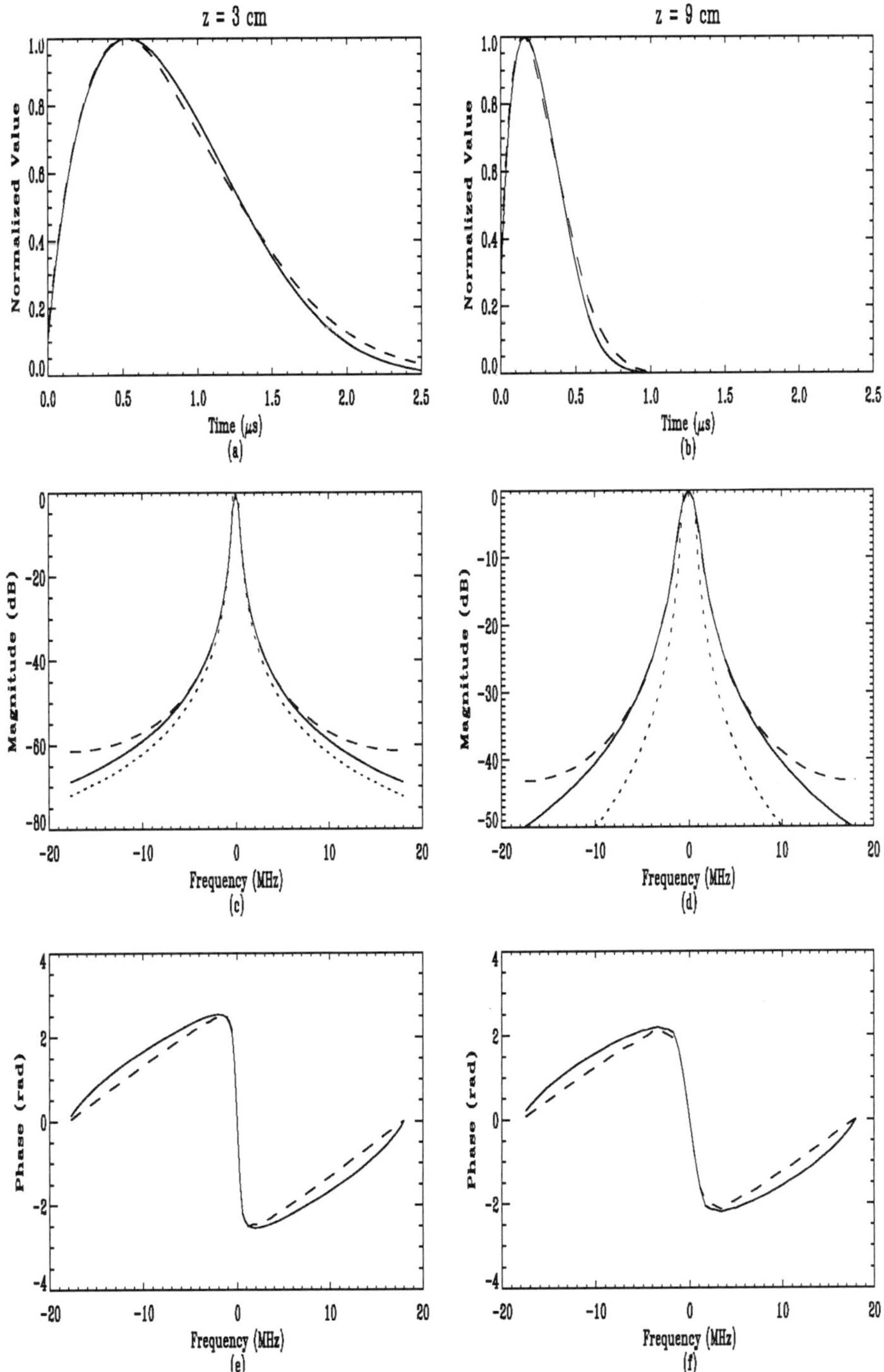

Figure 5.4. Two-way spatially averaged impulse responses for the Lommel (solid) and arccos (dashed) diffraction formulations: $b = a$.

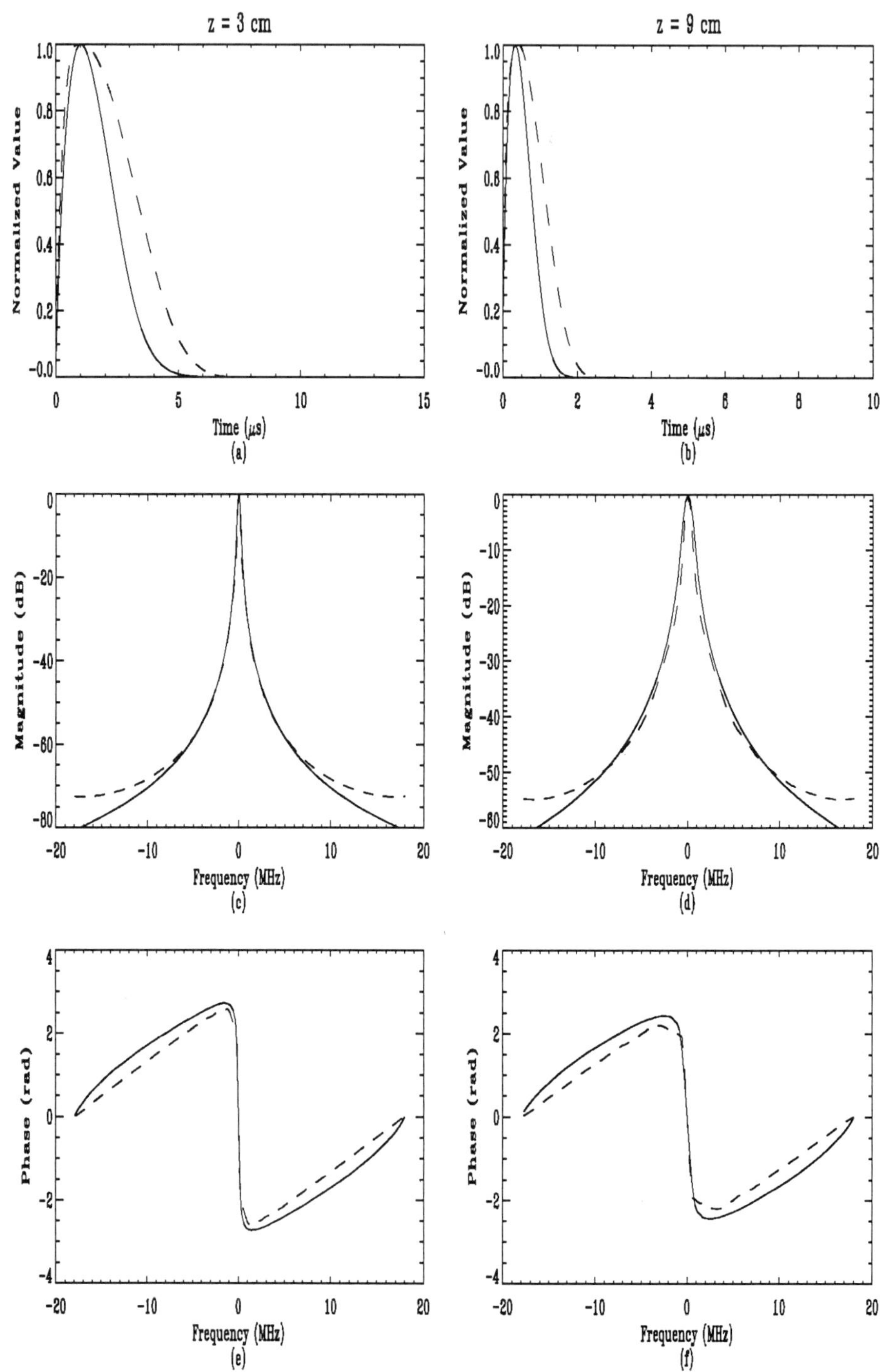

Figure 5.5. Two-way spatially averaged impulse responses for the Lommel (solid) and arccos (dashed) diffraction formulations: $b = 2a$.

6. COMPUTATIONAL CONSIDERATIONS

The five general computational issues discussed in Chapter 3 apply here; however, two caveats are needed. First, the time duration and, as a result, the number of samples of the two-way impulse response must be increased to account for the temporal effects of autoconvolution; this can be handled by sufficient zero-padding in the frequency domain. Second, frequency-domain windowing was not used with Eq. 5.25 to produce the plots in Figs. 5.2–5.5 because there are no dramatic discontinuities evident in the impulse responses. Thus, Gibb's phenomenon was not as pronounced as in the one-way case, and windowing was not required. This is not surprising because autoconvolution, in general, smoothes or averages out discontinuities [71, pp. 78–81].

Three new computational issues applicable only to spatially averaged autoconvolution diffraction require discussion. First, the arccos-based and Lommel-based impulse responses were normalized to unit amplitude because the scaling of $\langle h_2(z, t) \rangle_b$ remains an open question. The second computational consideration involves the assumption of minimum phase and demands more lengthy discussion. As noted earlier, the choice of sign in the minimum phase of Eq. 5.25 depends on the depth z and accounts for the time reversal of the focused arccos diffraction formulation when $z > A$. No such time reversal occurs in the unfocused case phase, and the sign is chosen based on the sign convention of the FFT. Incorrect choice of sign results in a time-reversed impulse response.

The minimum phase assumption also produced some unexpected results. Specifically, the spatially averaged autoconvolution impulse response is expected to be positive semi-definite because $h_1(\rho, z, t)$ is positive semi-definite. However, the phase response calculated using minimum phase sometimes caused the impulse response to be 180 degrees out of phase with the theoretically predicted value. This condition can be easily tested for and corrected algorithmically. Finally, the minimum phase assumption also produced unexplained transients in impulse responses computed for $b \ll a$. Considerable interpolation in the frequency domain and subsequent time-domain truncation is required to obtain meaningful autoconvolution impulse responses from Eq. 5.25 when $b \ll a$. How valid the minimum-phase assumption is in the focused case remains an open question.

7. CHAPTER SUMMARY

This chapter has shown that a set of equations derived by Wolf in 1951 for optical diffraction can be applied to spatially averaged autoconvolution diffraction corrections for both focused and unfocused piston transducers operating in pulsed mode. Wolf's expressions are based on Lommel's treatment of Fresnel diffraction and are magnitude-only expressions. Nonetheless, the expressions yield meaningful results in terms of describing attenuation due to diffraction for both focused and unfocused piston transducers. Minimum phase was assumed in order to estimate phase responses. Results computed with the Lommel-based expressions were compared to results obtained from numerical integration of the arccos diffraction formulation. Given the number of approximations and assumptions involved, the normalized results showed quite satisfactory agreement.

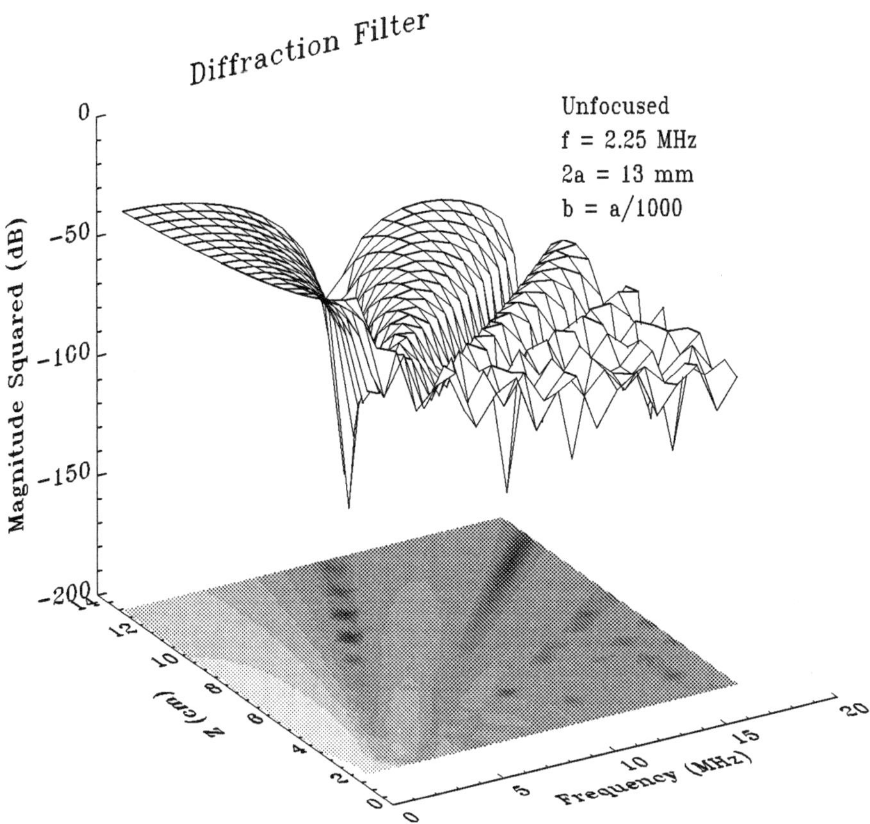

Figure 5.6.   Unfocused autoconvolution: $b = a/1000$.

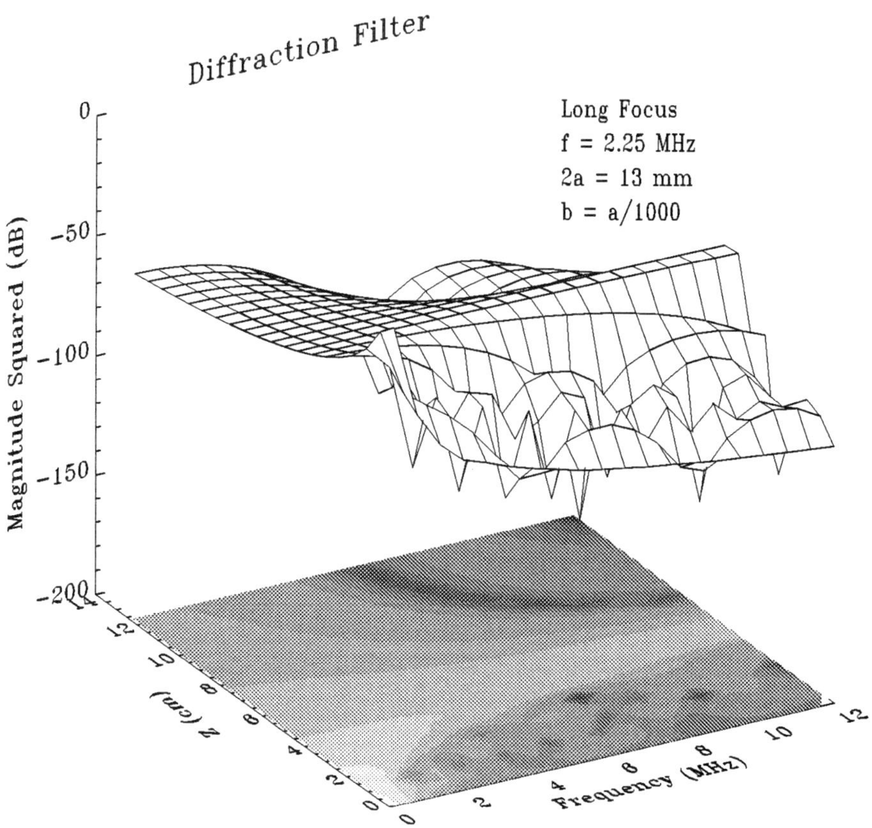

Figure 5.7.　　Long focus autoconvolution: $b = a/1000$.

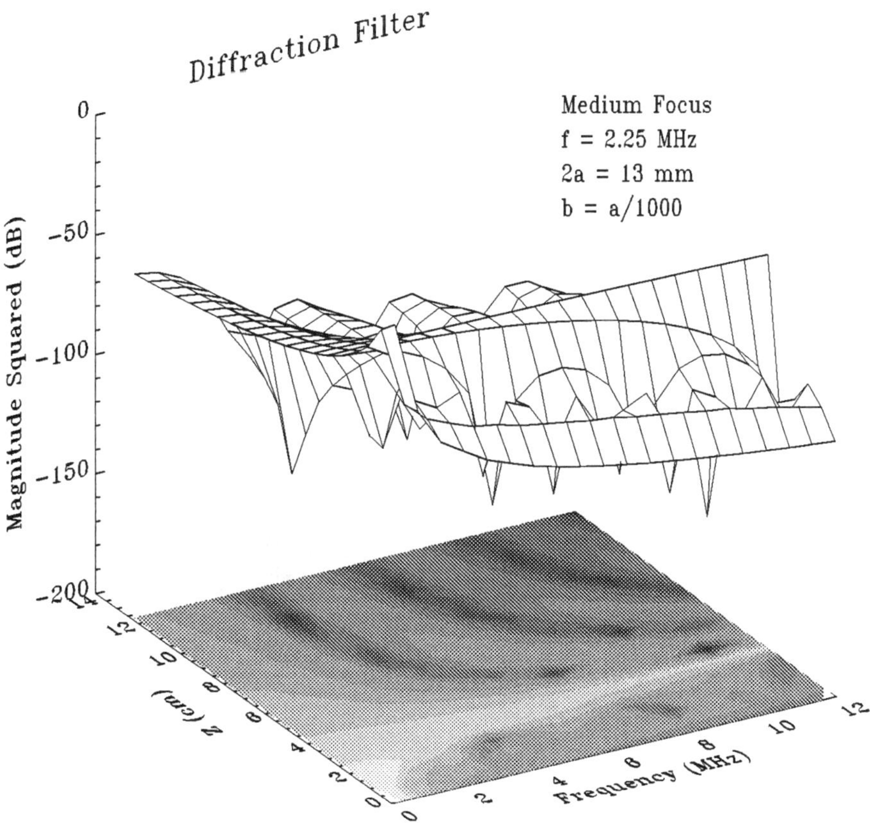

Figure 5.8.   Medium focus autoconvolution: $b = a/1000$.

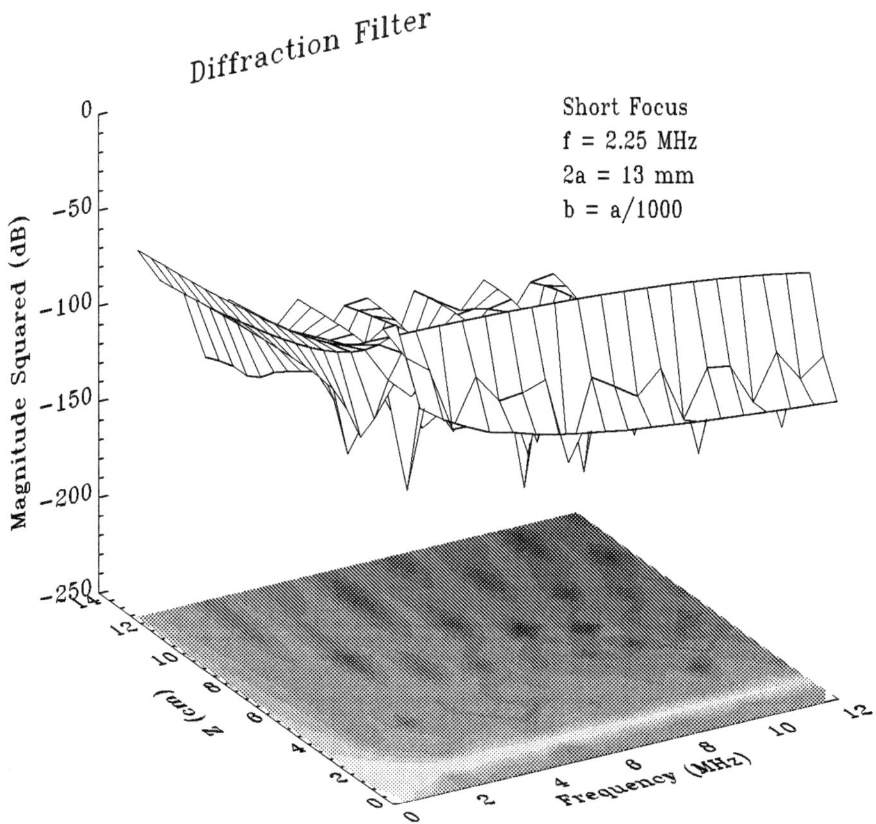

Figure 5.9.   Short focus autoconvolution: $b = a/1000$.

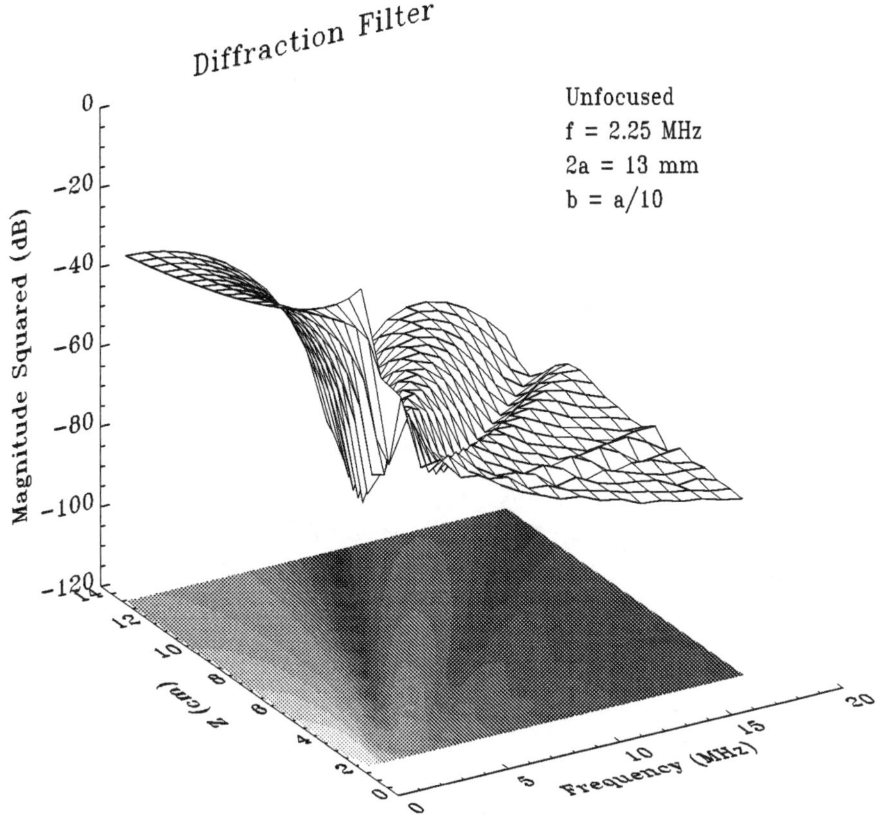

Figure 5.10.   Unfocused autoconvolution: $b = a/10$.

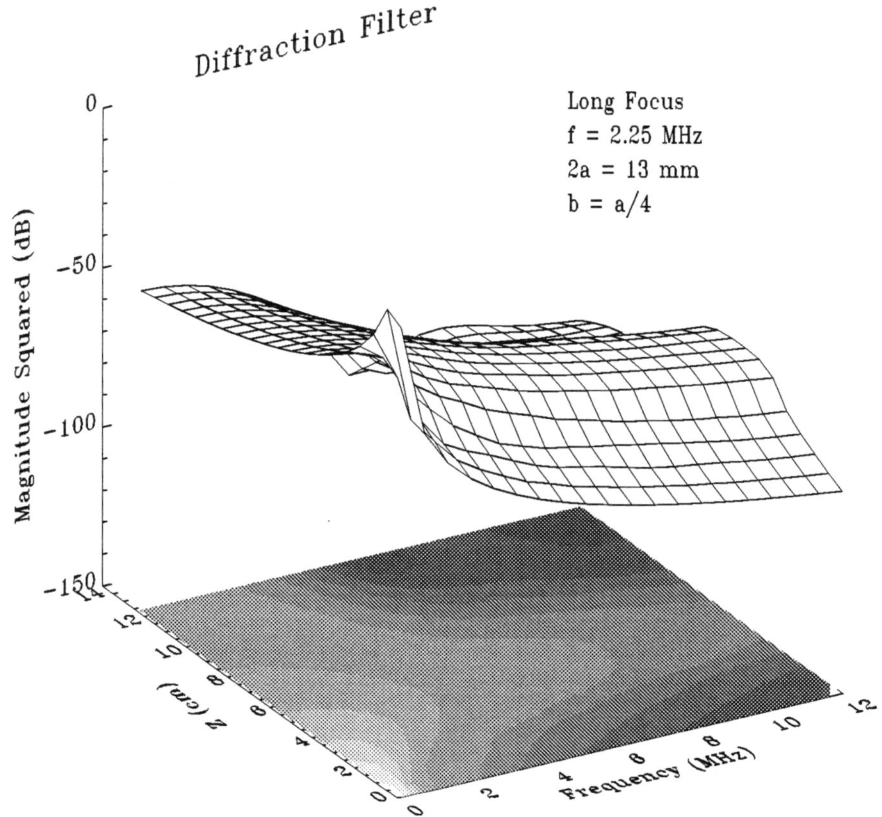

Figure 5.11.   Long focus autoconvolution: $b = a/4$.

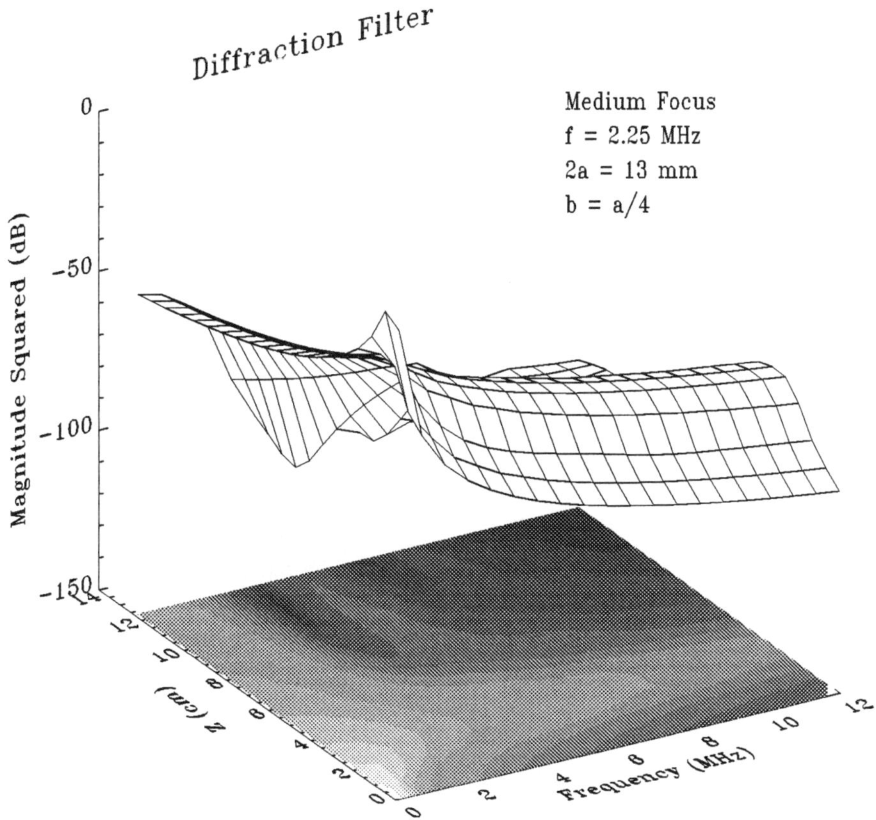

Figure 5.12.   Medium focus autoconvolution: $b = a/4$.

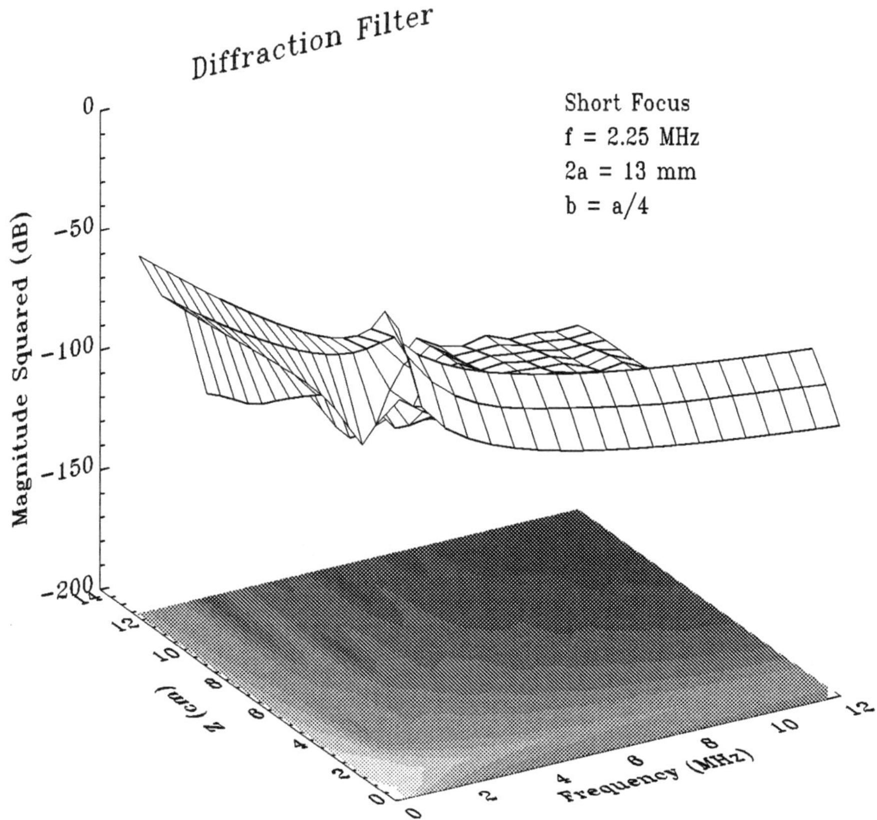

Figure 5.13.   Short focus autoconvolution: $b = a/4$.

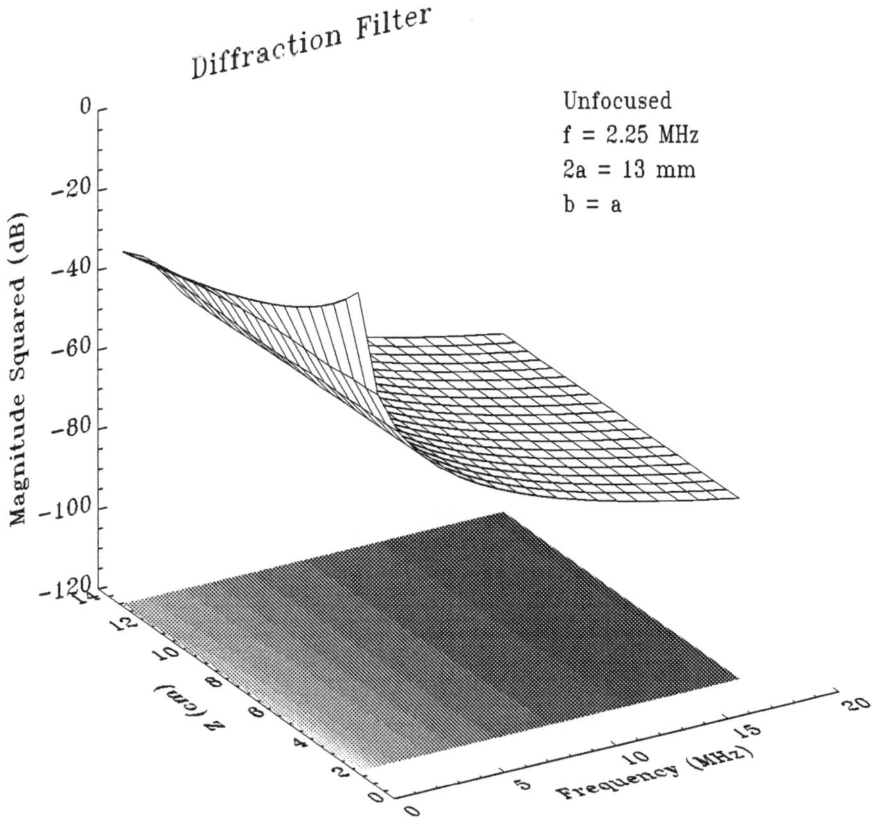

Figure 5.14.   Unfocused autoconvolution: $b = a$.

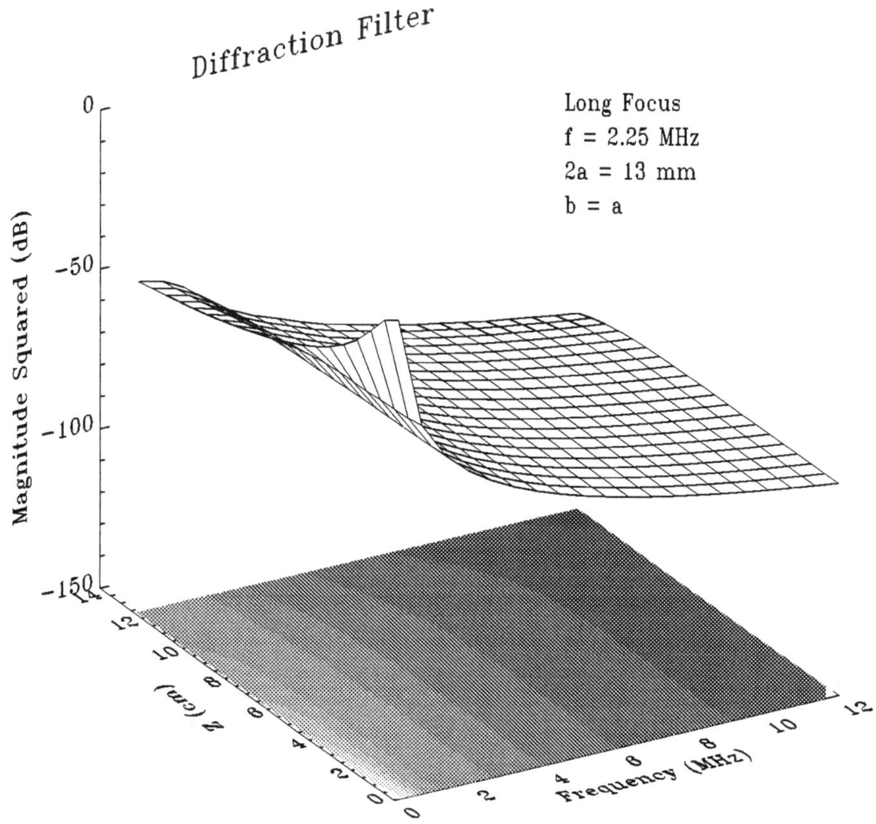

Figure 5.15.   Long focus autoconvolution: $b = a$.

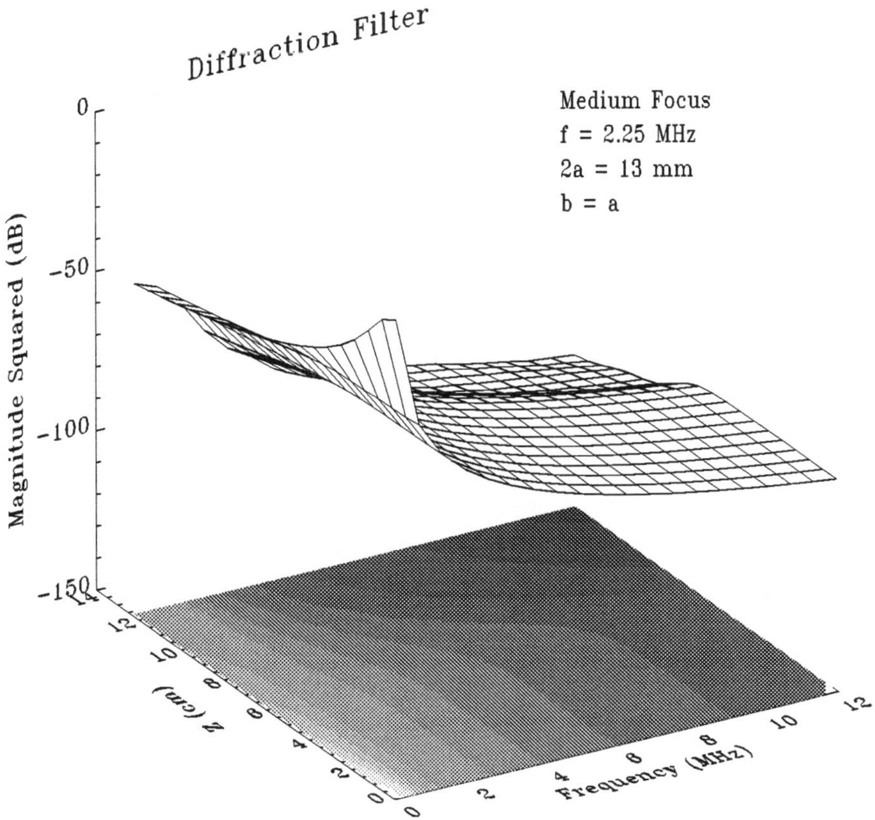

Figure 5.16.   Medium focus autoconvolution: $b = a$.

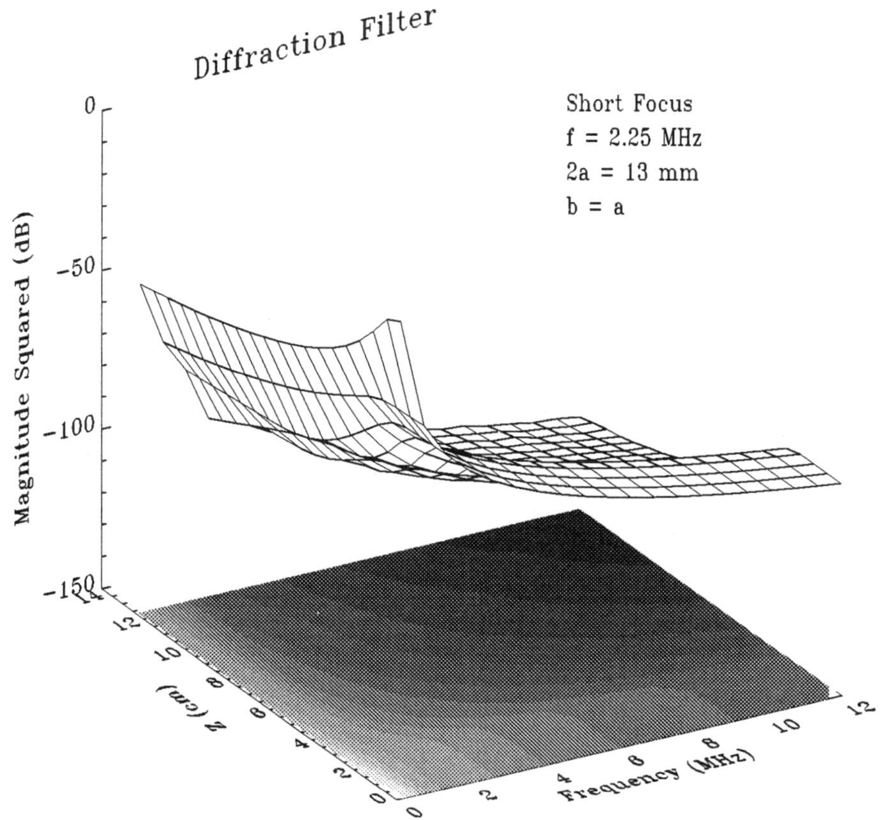

Figure 5.17. Short focus autoconvolution: $b = a$.

Chapter 6

EXPERIMENTAL INVESTIGATION

The discussion up to this point has focused on proposing and verifying a new theoretical perspective of spatially averaged diffraction correction. In this chapter, an aspect of the proposed theory is investigated experimentally. Specifically, autoconvolution diffraction corrections will be implemented with time-varying filters, and diffraction-corrected B-mode images will be reconstructed using a short-time Fourier technique. The raw and diffraction-corrected images will be compared in a qualitative sense only. Differences between raw and corrected RF data will be quantitatively analyzed via spectral centroids which were introduced Chapter 1.

At this point it is necessary to reiterate three points discussed in Chapter 1. First, the experiments were not designed to verify the theory in any authoritative fashion. Rather, they were designed to gauge the feasibility of the proposed autoconvolution diffraction correction. Second, the diffraction-corrected images are a significant feature of this work. Although the differences between the raw and diffraction-corrected images are subtle, they reveal that diffraction correction has a more pronounced effect on RF data than it does on envelope-detected data. Finally, and most importantly, the experiments are not to be considered, in any way, clinical validation of the proposed diffraction corrections.

It is also important to note that the autoconvolution diffraction corrections are based on Eq. 5.25 which was verified in terms of magnitude and phase for unfocused piston transducers only. Hence, both magnitude and phase responses were computed and applied for unfocused corrections while only magnitude responses were computed and applied for focused corrections.

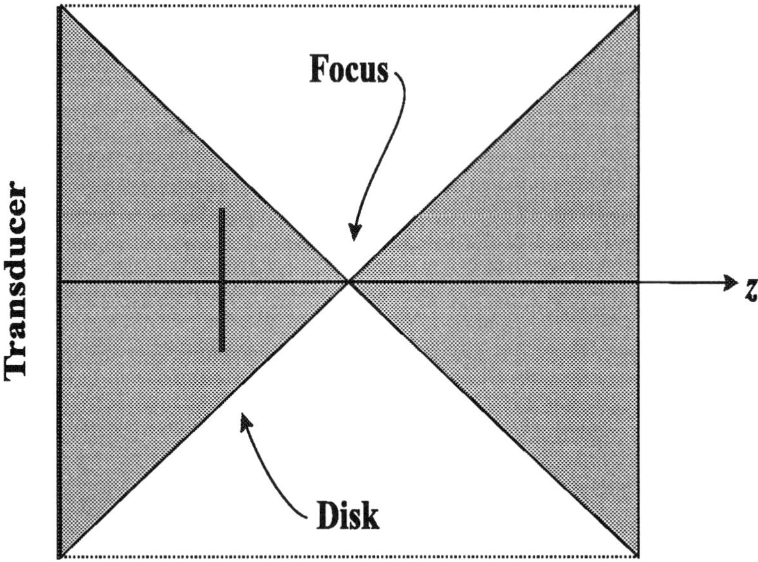

Figure 6.1.   Reflecting plate in main beam.

1.   A COMPUTATIONAL CONSIDERATION

Figs. 5.6–5.17 in the previous chapter showed that the nature of a given two-way diffraction filter depends heavily on the value of b, the radius of the reflecting disk, used to calculate the diffraction filter. This, coupled with the fact that b is theoretically unrestricted, raises a new computational issue. That is, to the extent that Eqs. 5.20–5.22 and 5.25 accurately model reality, what b should be used to compute the diffraction filter?

An imprecise but intuitively appealing answer can be had if we consider Fig. 6.1 in terms of conservation of energy. The figure shows a reflecting disk of finite extent located in the beam of a focused transducer. Since no attenuation mechanism has been included in the theory, the total acoustic energy is constant for all z, but the energy density is not. Loosely speaking, the energy gets more concentrated at the focus.

Experience confirms this assertion. Consider insonifying a large ($b \gg a$) flat disk with a piston transducer. The amplitude of the received pulses is maximized at the focus and decreases on either side of the focus. This behavior could be accounted for by considering goniemetric aspects of the scattering phenomenon, but this would be a very difficult undertaking.

Instead, we borrow a concept from the ultrasonic literature [83, 28] and propose the use of an *effective backscattering cross-section* to account for the reflection phenomenon just described. The radar community has successfully resorted to such *ad hoc* and empirical methods involving a variety of effective parameters when faced with intractable mathematical problems. Indeed, it can be argued that a cornerstone of modern radar theory, the radar cross-section (RCS), is ultimately an *ad hoc* parameter [12, 85].

For the purposes of this work, the effective backscattering cross-section (BCS) $\hat{\sigma}_{BS}$ is defined

$$\hat{\sigma}_{BS} = \pi b^2 = \pi (K_{eff}a)^2. \tag{6.1}$$

Thus, $b = K_{eff}a$, where K_{eff} is the parameter used to control the effective BCS and $0 \leq K_{eff} < \infty$. Incorporating a, the radius of the transducer, in this admittedly *ad hoc* definition of effective BCS allows us to relate the effective BCS of the flat plate to the diameter of the transducer which, in turn, will allow us to relate effective BCS to, say, 3 dB beamwidth.

In this work, K_{eff} was set to approximately 0.5 for unfocused transducers and to K_F for focused transducers where K_F is the focusing factor. The focusing factor is approximately 0.2, 0.5, and 0.8 for short-, medium-, and long-focus transducers, respectively [53, 69]. In practice, K_{eff} may have to be varied as a function of depth z.

2. EQUIPMENT AND PROCESSING

Unfocused diffraction corrections based on Eq. 5.25 and focused diffraction corrections based on the magnitude of Eq. 5.25 were applied to RF echo data obtained from a variety of piston transducers operating in pulsed mode. The RF data were obtained using the equipment shown in Fig. 6.2. A Panametrics model 5052PR pulser/receiver provided transducer excitation and initial amplification of the RF echo data. A stepper motor (not shown) was used to move the transducer. RF echo data of interest were selected with a Panametrics model 5052G gate. Additional amplification was provided by a RITEC model BR-640 broadband receiver, and the RF data was digitized to 8-bit resolution by a Data Precision 6100B universal waveform analyzer. The RF data were subsequently downloaded to a personal computer (PC).

All subsequent signal processing was done on the PC. Specifically, a diffraction filter was calculated using Eq. 5.25 for the bandwidth of interest. Next, the filter was normalized by its maximum value, and the corresponding inverse filter was calculated. The resulting diffraction correction was implemented as a time-varying filter via a short-time

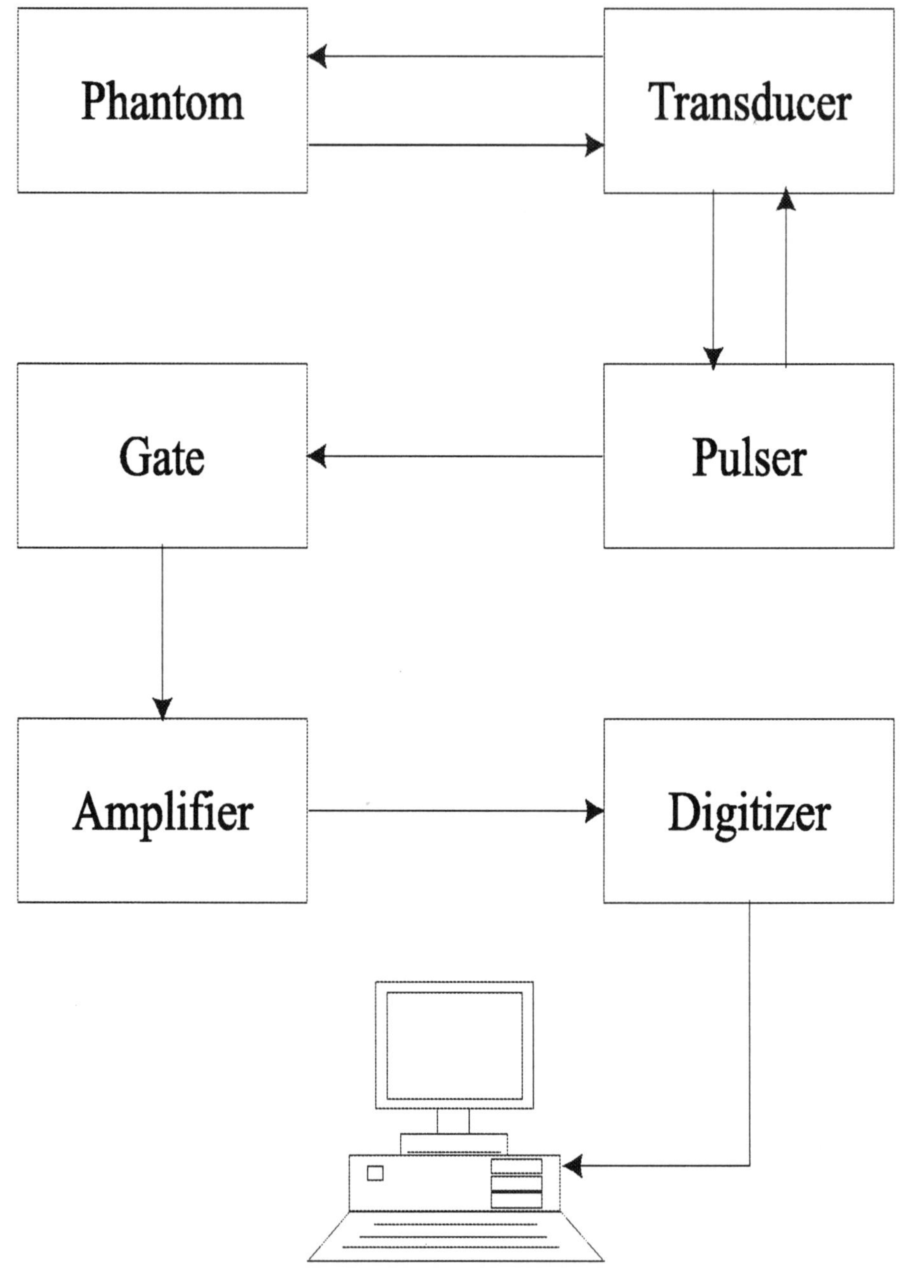

Figure 6.2.   Experimental set-up.

Fourier technique known as the *weighted overlap-add* (WOLA) method [22]. Fig. 6.3 depicts the WOLA processing used in this work.

The following short-time Fourier parameters were used: hop length of 4 samples, 256 filter coefficients from DC to the Nyquist frequency, and a window length of 31 samples zero-padded to 256 samples. See Reference [74] for parameter definitions. A Hamming window was used for analysis, while a rectangular window was used for synthesis. The bandwidth of interest was defined as $f_c \pm f_c/2$, where f_c is the center frequency of the transducer. The RF sampling rate of the Data Precision 6100B was set at 40 nanoseconds and 20 nanoseconds for $f_c \leq 2.25$ MHz and $f_c > 2.25$ MHz, respectively. The sampling rate was later reduced by a factor of one half in software; that is, every other sample was discarded. In each experiment except the wire experiment, the distance between scans or lateral sampling rate was 0.3 mm, and no averaging was done in either the axial or lateral direction.

3. EXPERIMENTS, IMAGES, AND CENTROIDS

Results obtained from ten imaging experiments along with relevant imaging parameters are shown in Figs. 6.4–6.28 which are grouped at the end of the chapter for convenience. Figures containing gray-scale images obtained with *unfocused* transducers are followed by two figures which indicate the magnitude and phase filtering done to obtain the diffraction-corrected data. Figures containing gray-scale images obtained with *focused* transducers are followed by one figure which indicates the magnitude filtering done to obtain the diffraction-corrected data.

The format of the figures containing gray-scale images is as follows. The gray-scale images labeled (a) in the figures show B-mode images constructed from raw RF data. The images in Fig. 6.4(a) and Fig. 6.7(a) are from a specimen of unfixed cancerous human breast tissue in water on a sponge (100 A-lines). The images in Fig. 6.10(a), Fig. 6.13(a), Fig. 6.15(a), and Fig. 6.18(a) are from a specimen of unfixed pig liver in water on a sponge (50 A-lines). The images in Fig. 6.20(a), Fig. 6.22(a), and Fig. 6.24(a) are from an ATS Laboratories Model 539 attenuating phantom (60 A-lines). The wire-target data shown in Fig. 6.26(a) are from a standard AIUM wire phantom (24 A-lines with a 0.8 mm lateral sampling rate). The gray-scale images labeled (b) show B-mode images constructed from diffraction-corrected RF data. No clinical data on the tissue or liver samples is provided because the experiments were not designed to demonstrate or validate the clinical efficacy of the proposed diffraction corrections.

WOLA Diffraction Correction of *i*-th A-line

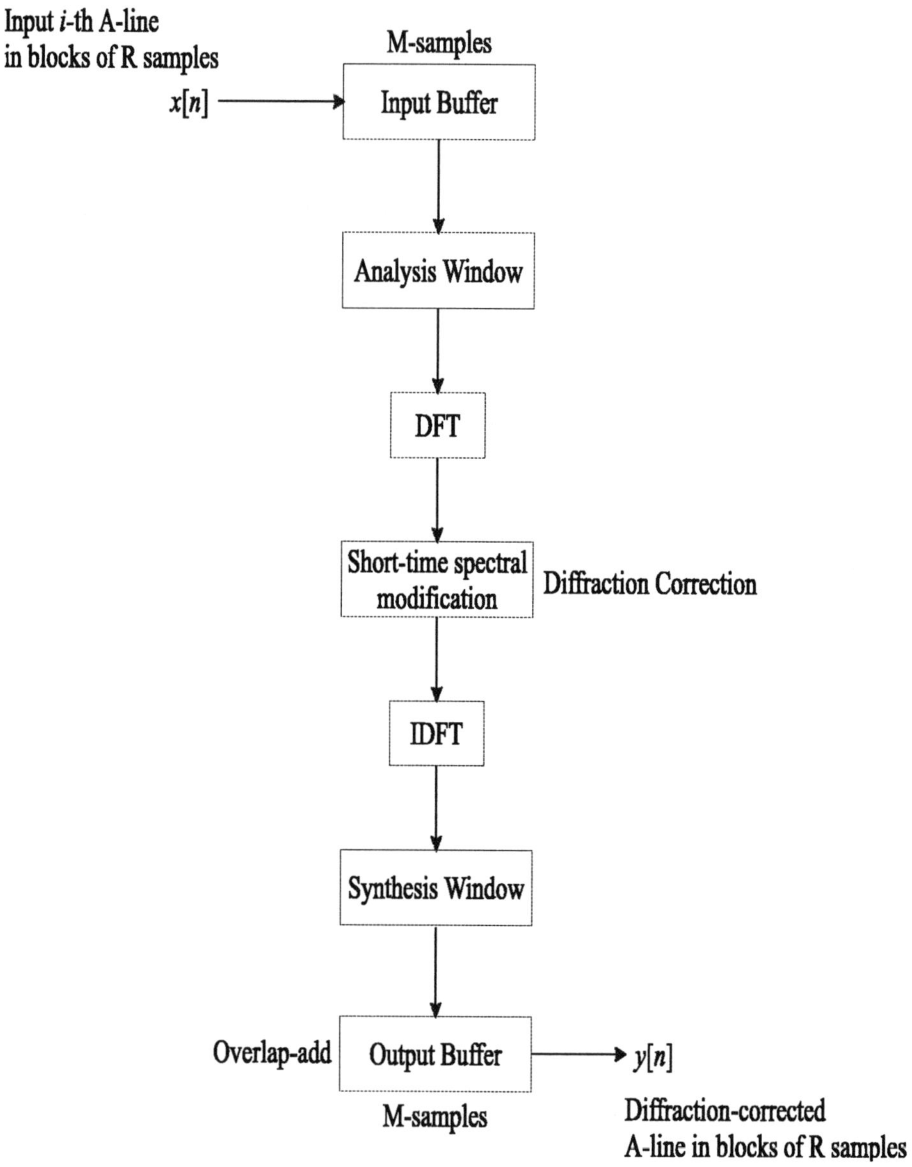

Figure 6.3.   Short-time Fourier processing of A-lines.

The plots labeled (c) and (d) in the figures containing gray-scale images are the FFT's of the middle A-line from the raw and corrected RF data, respectively. The FFT data were smoothed *after* log scaling to dB; hence, the maximum value is not 0 dB. The centroid plots labeled (e) and (f) in the figures with gray-scale images were made by calculating, via short-time Fourier transformation, the spectral centroid of the middle A-line from the raw and corrected RF data, respectively. The centroid plots are annotated with the mean μ and standard deviation σ, in Hz, computed from the respective centroid record. Recall from Chapter 1 that the spectral centroid is a measure of the mean frequency of a signal.

The figures immediately following the figures containing gray-scale images show the magnitude response and, for unfocused transducers, the phase response of the diffraction correction. The responses are plotted as a function of frequency in the column on the left, while they are plotted as a function of depth in the column on the right. These plots give an indication of the filtering that was done. Clearly, Gibb's phenomenon was an issue because of the sharp cut-offs exhibited by the filters. The resultant ringing was ameliorated by two factors, Hamming analysis windows reduced ringing, and the repeated overlap and add operations, in a sense, averaged or smoothed out the remaining ringing.

4.   DISCUSSION OF RESULTS

Diffraction correction affected the data in three ways. First, it performed a kind of depth-dependent time-gain correction which can be seen by comparing the raw and filtered gray-scale images. The effect can be seen quite clearly in Fig. 6.13 and Fig. 6.20. Second, the diffraction correction is, in a spectral sense, a depth-dependent high-frequency amplifier. Compare the raw and filtered gray-scale images. Qualitatively speaking, the diffraction-corrected images look crisper or sharper than the raw images. Additionally, compare the raw and filtered FFT's shown in the figures containing gray-scale images. Clearly, the high-frequency information contained in the raw data has a lower magnitude than that contained in the filtered or diffraction corrected data. Third, diffraction correction appears to reduce the variance of the spectral centroid. The variance reduction is more noticeable for focused transducers than for unfocused, but a quantifiable reduction has been achieved for all cases except the one shown in Fig. 6.10.

5.   CHAPTER SUMMARY

This chapter presented a brief experimental investigation of the proposed diffraction theory developed in previous chapters. Specifically, autoconvolution diffraction corrections were implemented with time-varying filters, and diffraction-corrected B-mode images were reconstructed using a short-time Fourier analysis/synthesis algorithm known as the weighted overlap-add (WOLA) method. The raw and diffraction-corrected images were compared only qualitatively. Differences between raw and corrected RF data were quantitatively analyzed via spectral centroids, a measure of average frequency. The experimental results prove that, at worst, the proposed diffraction correction does no harm and, at best, it removes spectral bias and appears to improve image quality.

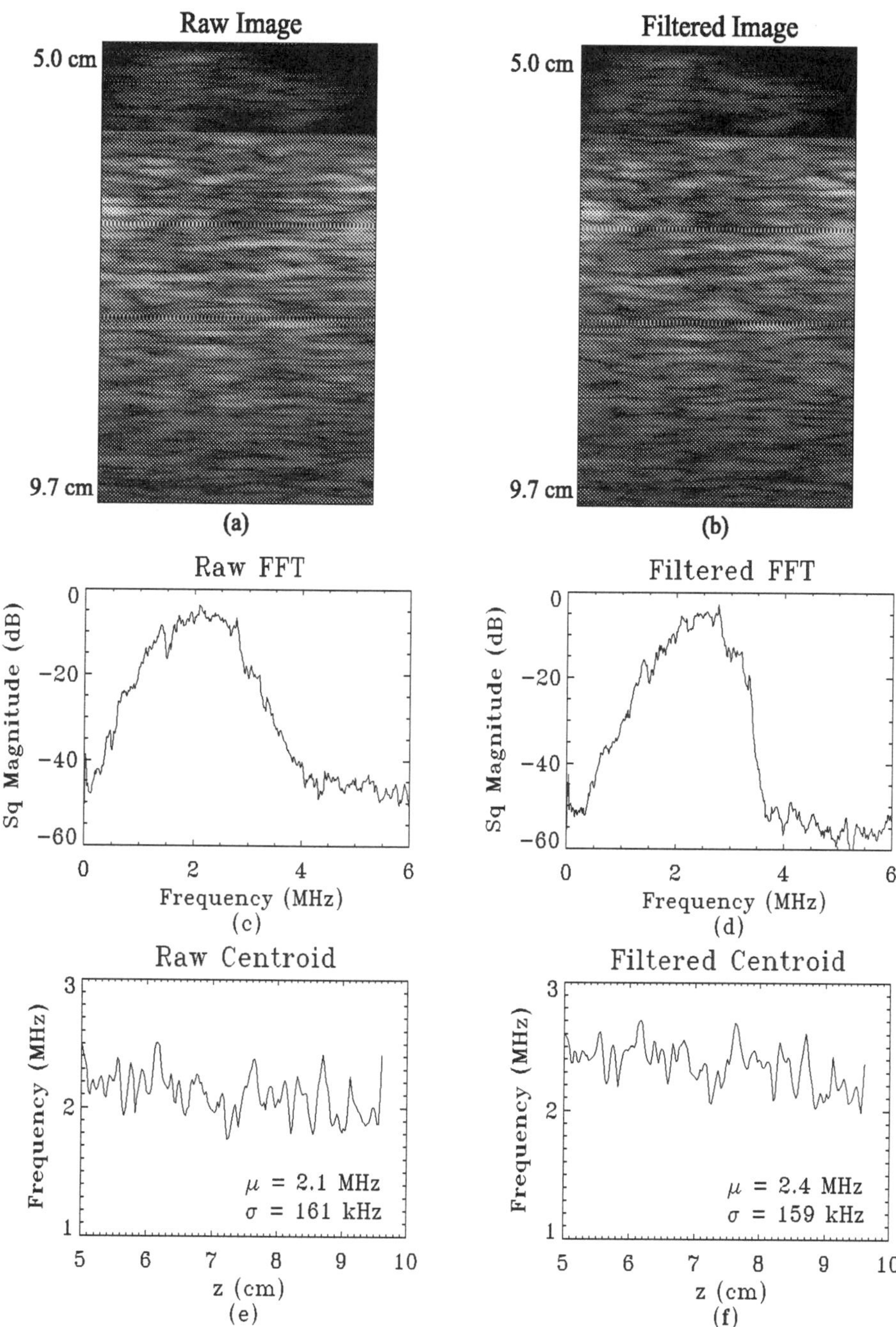

Figure 6.4.   Breast on sponge — 2.25 MHz unfocused ($2a = 13$ mm).

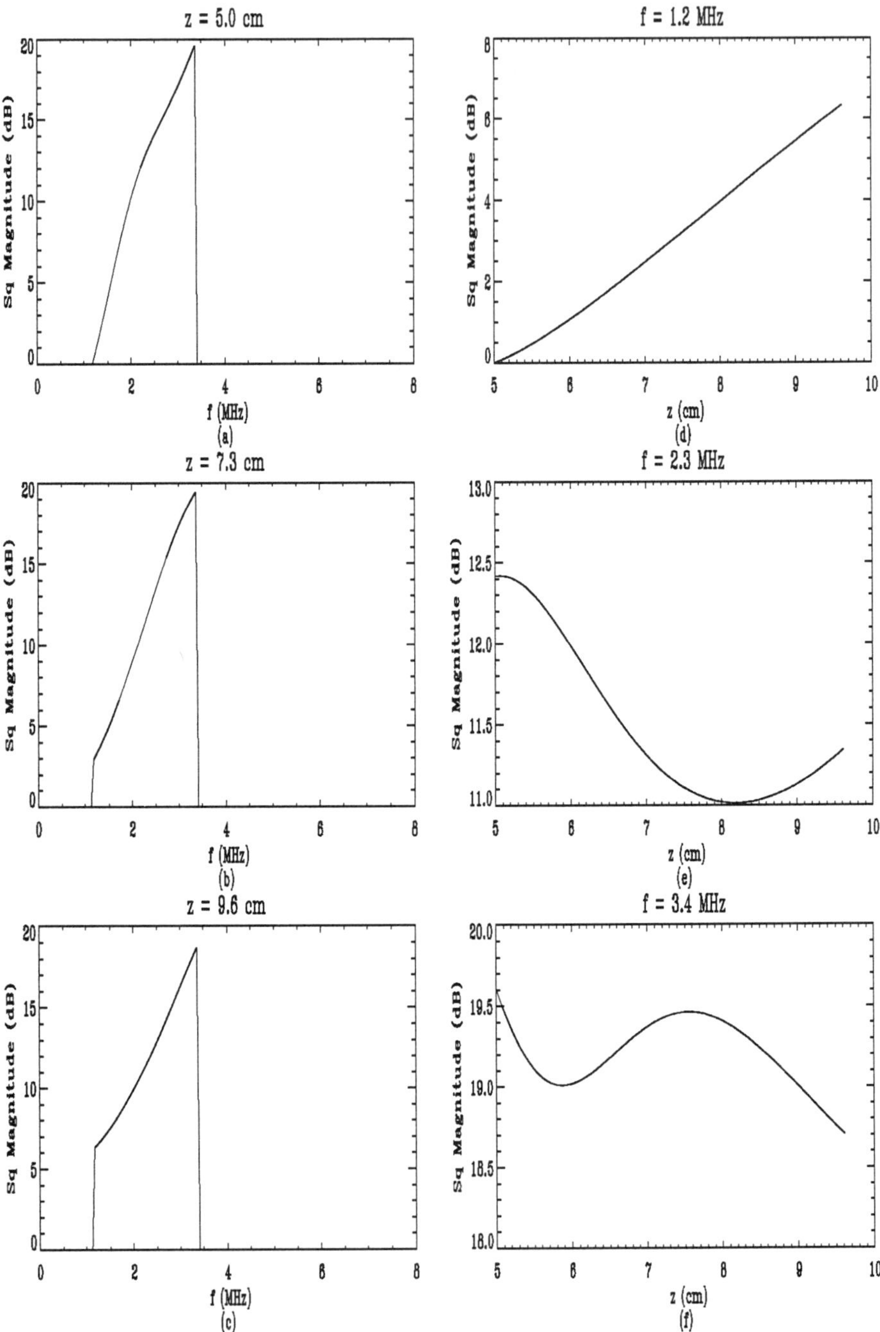

Figure 6.5.   Magnitude response — 2.25 MHz unfocused ($2a = 13$ mm).

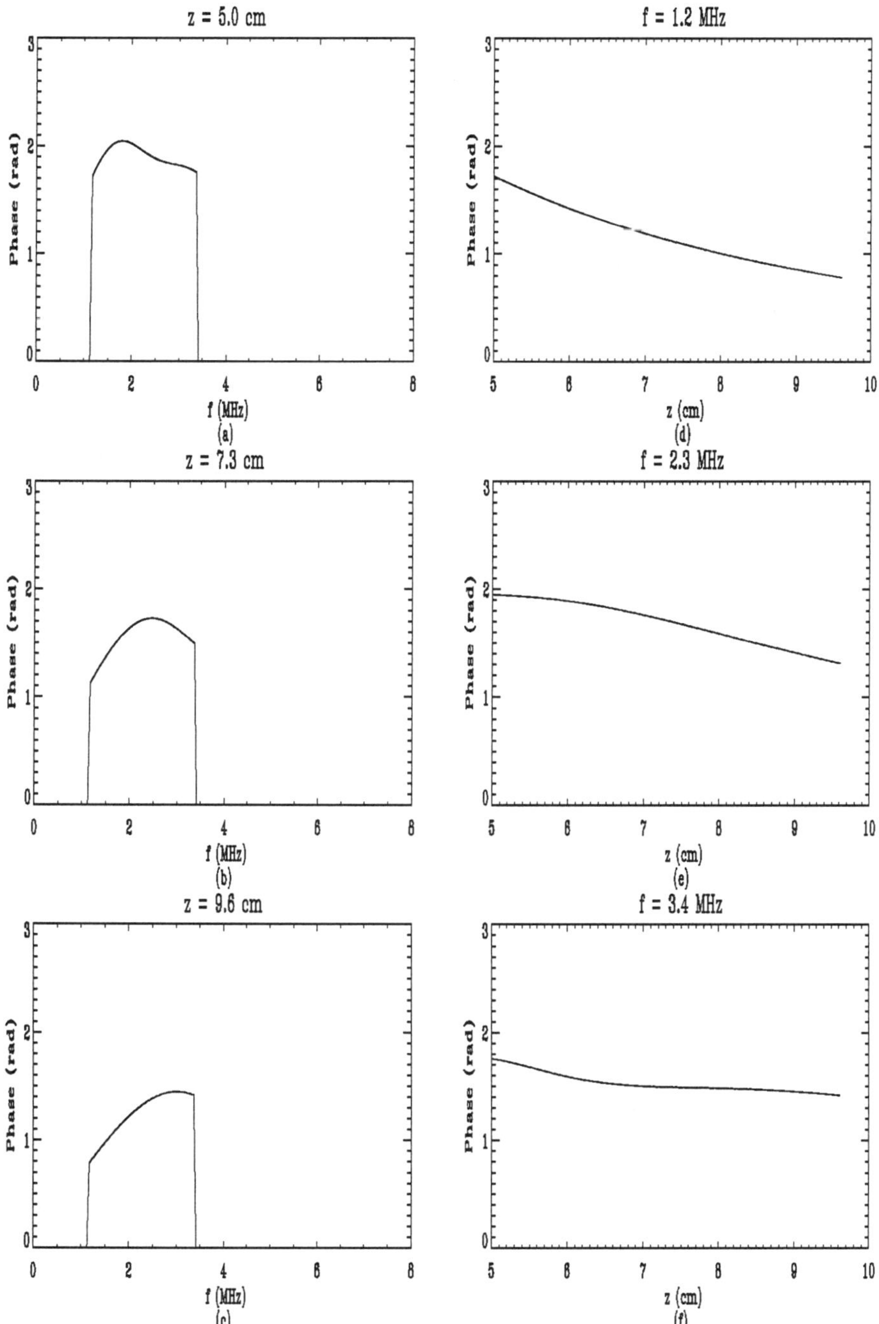

Figure 6.6.   Phase response — 2.25 MHz unfocused ($2a = 13$ mm).

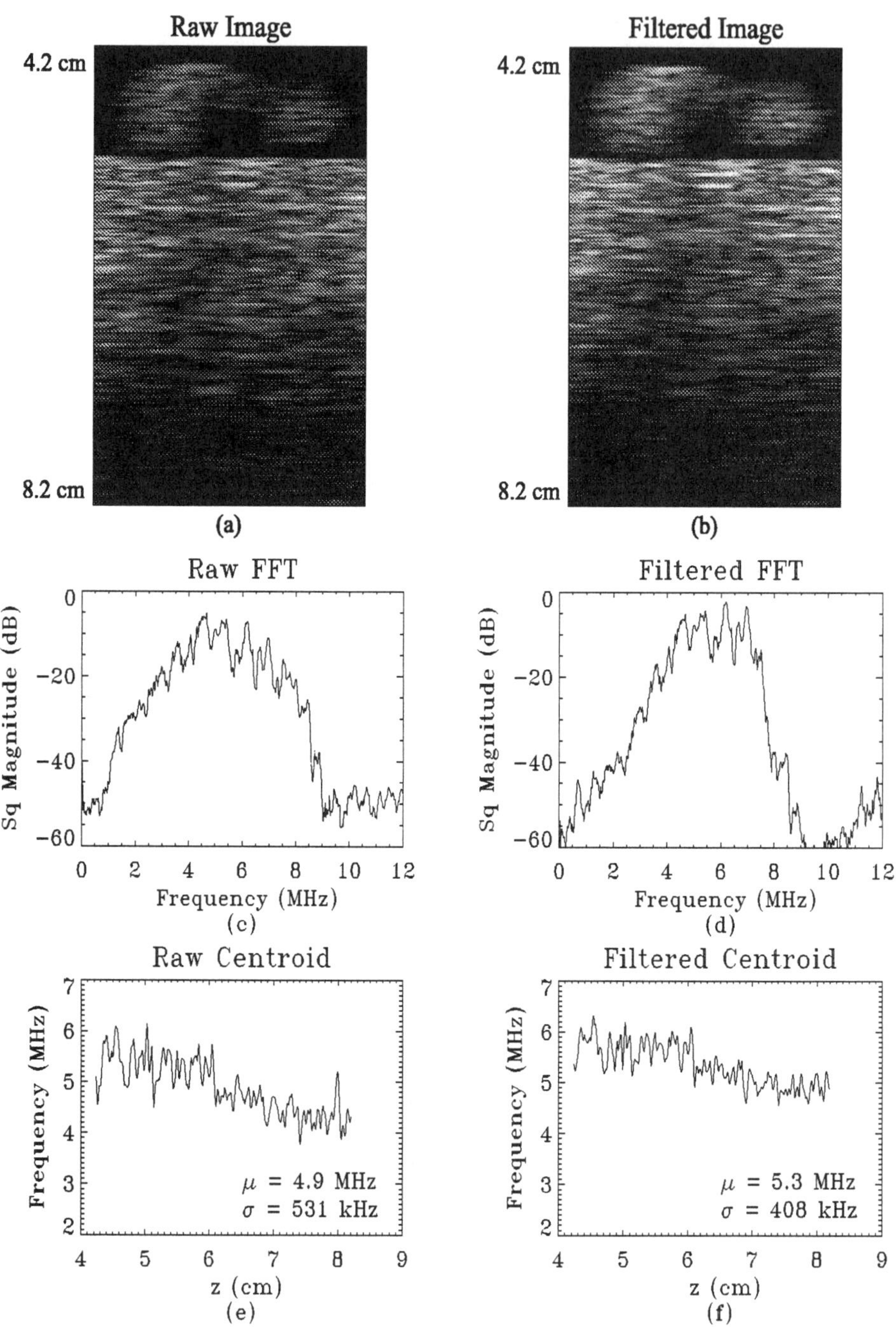

Figure 6.7.   Breast on sponge — 5.0 MHz unfocused ($2a = 9.53$ mm).

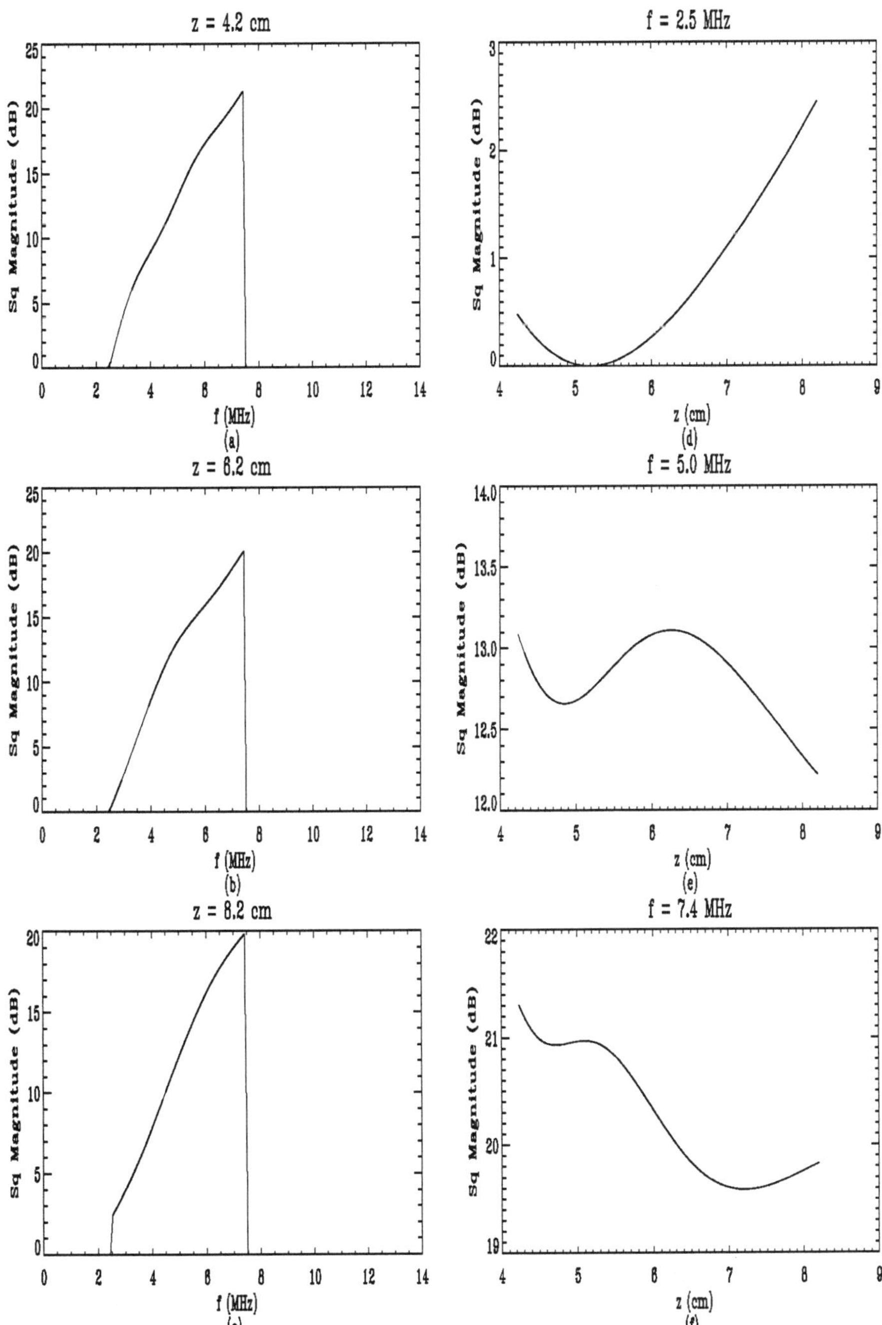

Figure 6.8.   Magnitude response — 5.0 MHz unfocused ($2a = 9.53$ mm).

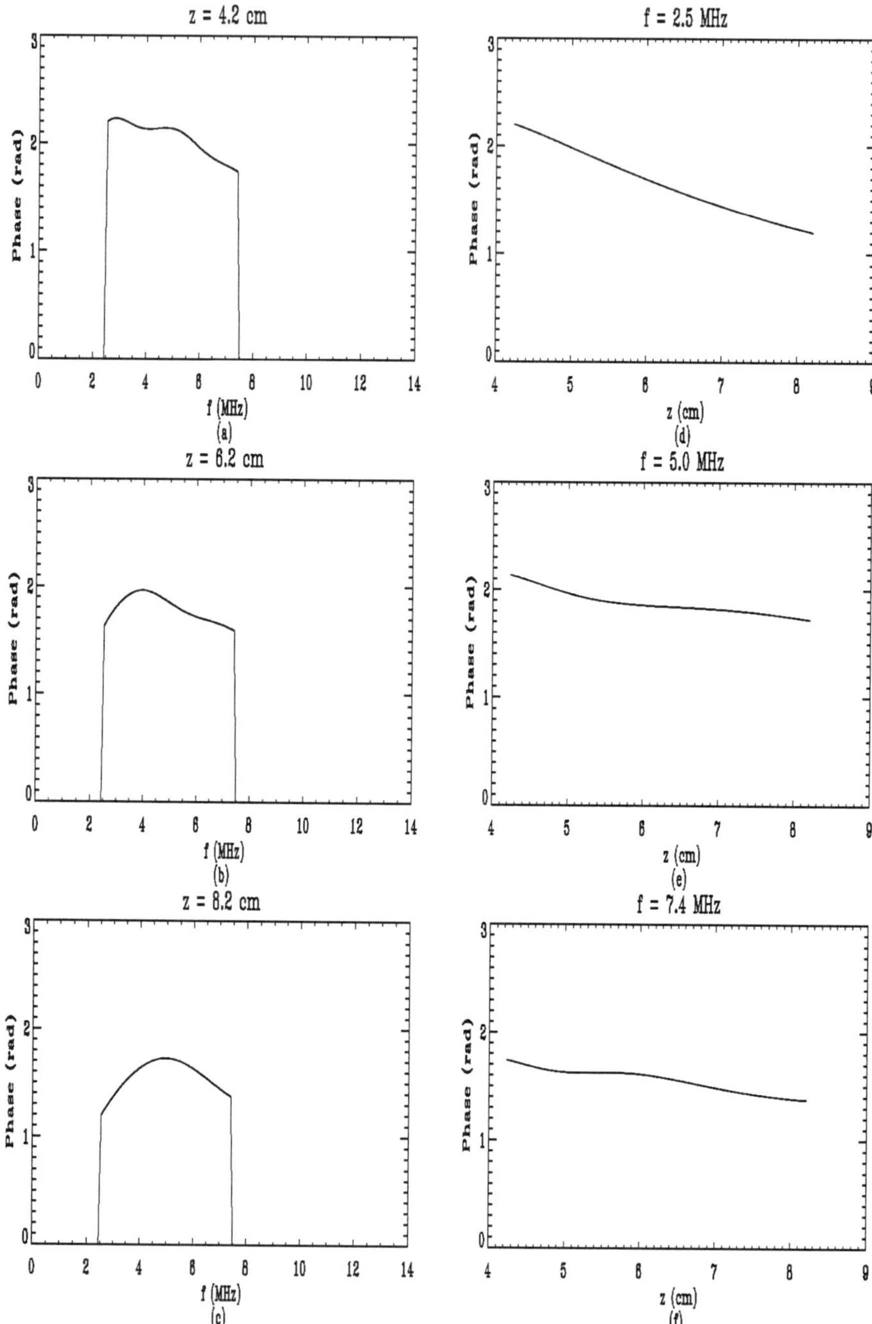

Figure 6.9.   Phase response — 5.0 MHz unfocused ($2a = 9.53$ mm).

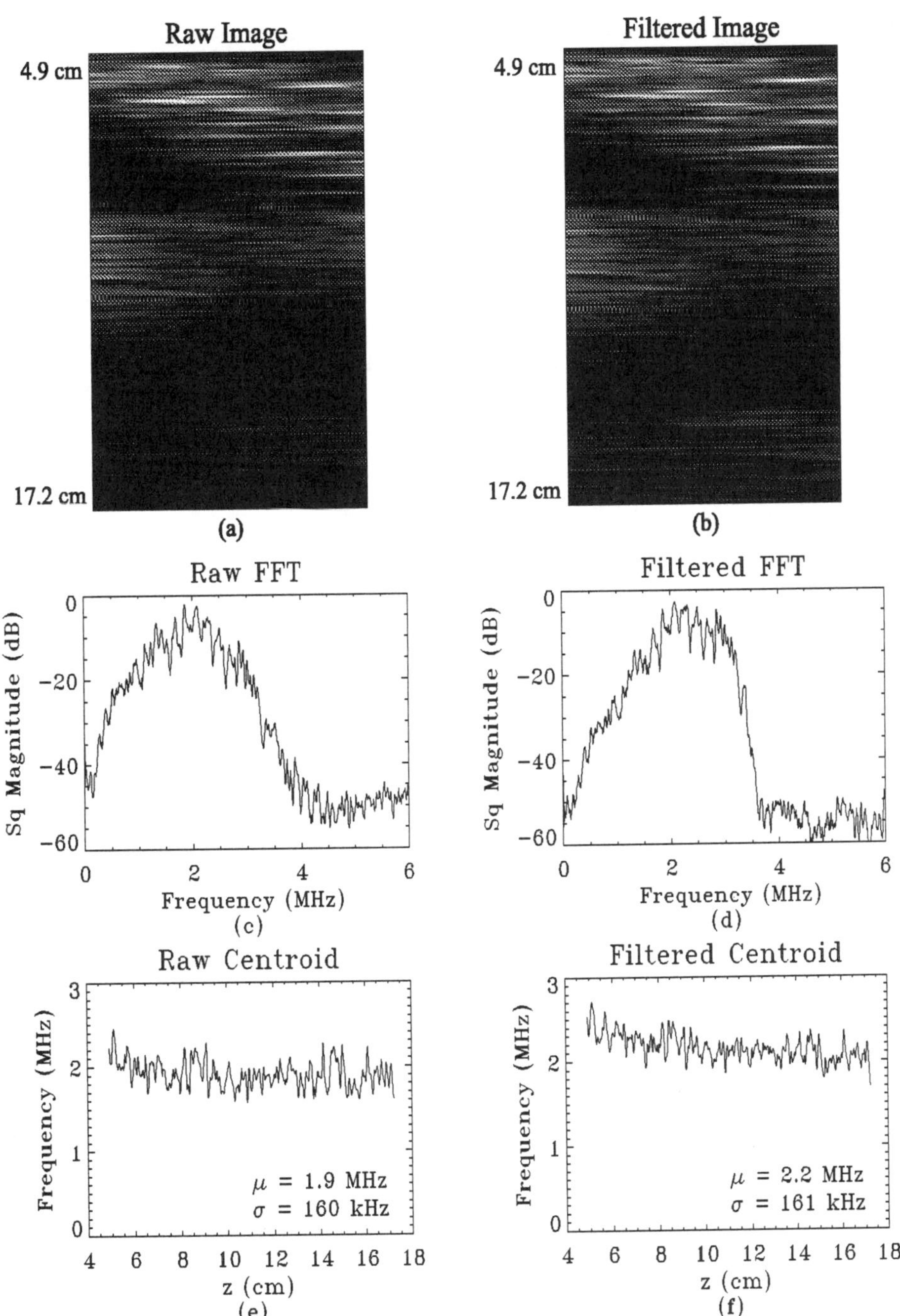

μ = 1.9 MHz
σ = 160 kHz

μ = 2.2 MHz
σ = 161 kHz

Figure 6.10.   Pig liver on sponge — 2.25 MHz unfocused ($2a = 13$ mm).

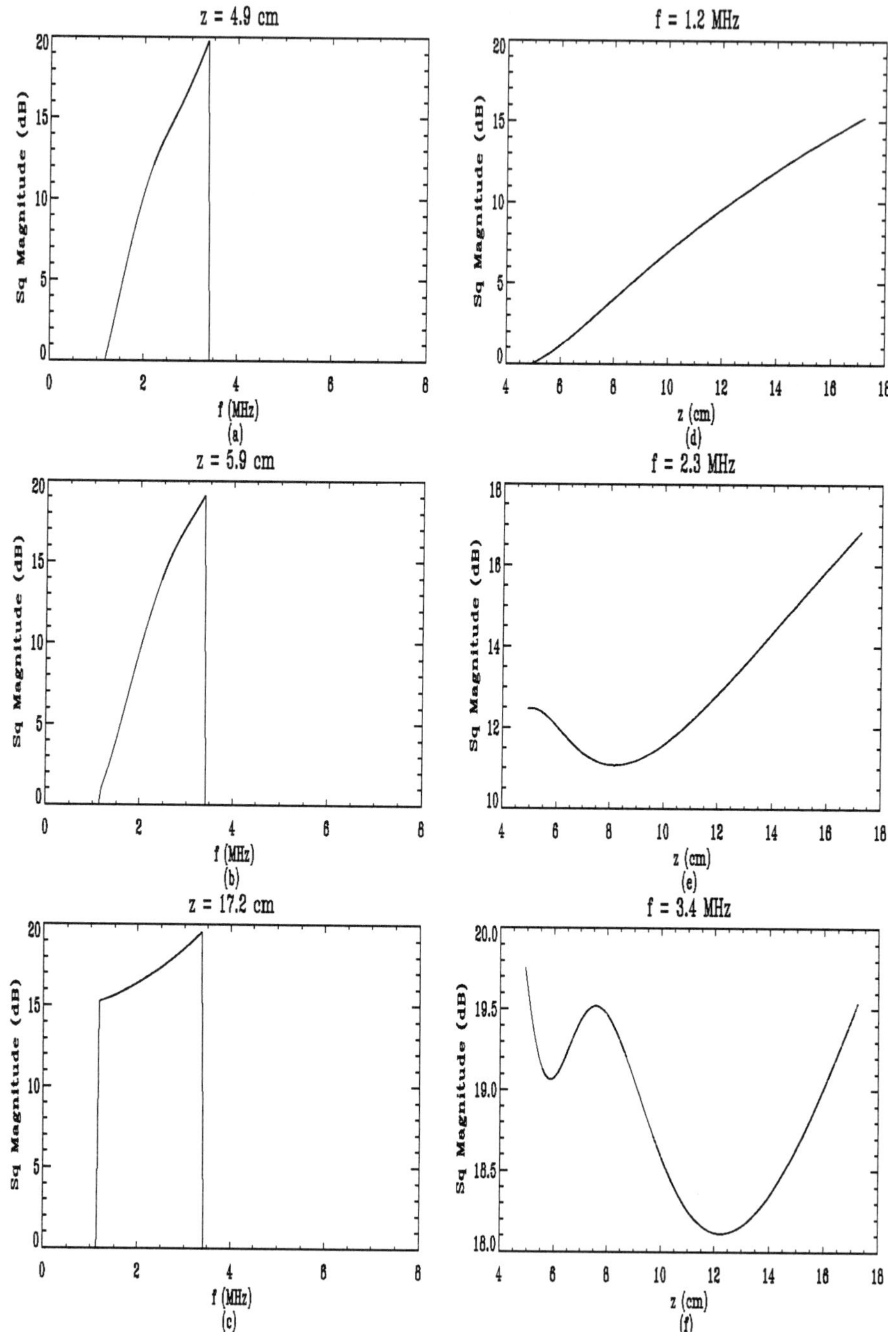

Figure 6.11.   Magnitude response — 2.25 MHz unfocused ($2a = 13$ mm).

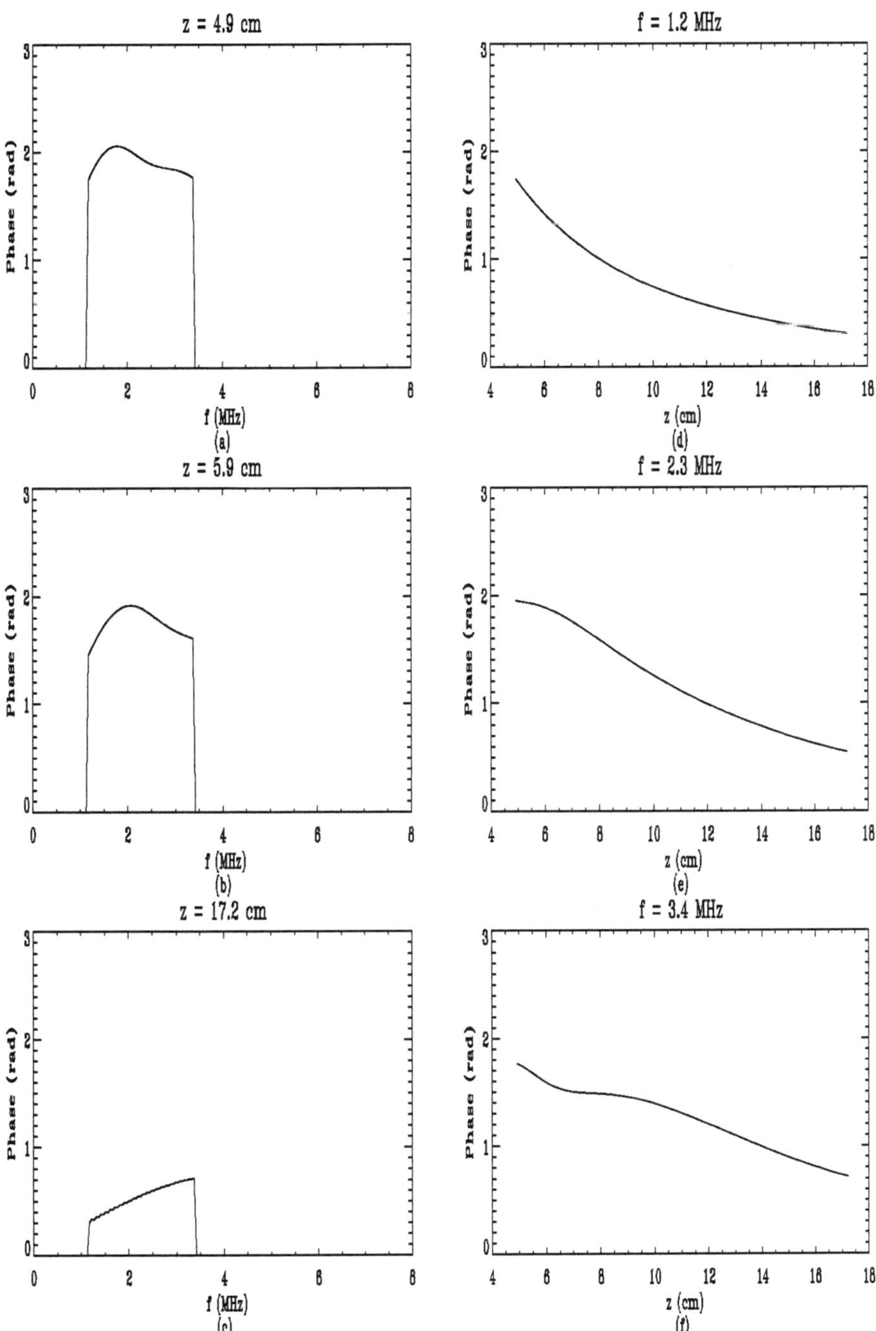

Figure 6.12.   Phase response — 2.25 MHz unfocused ($2a = 13$ mm).

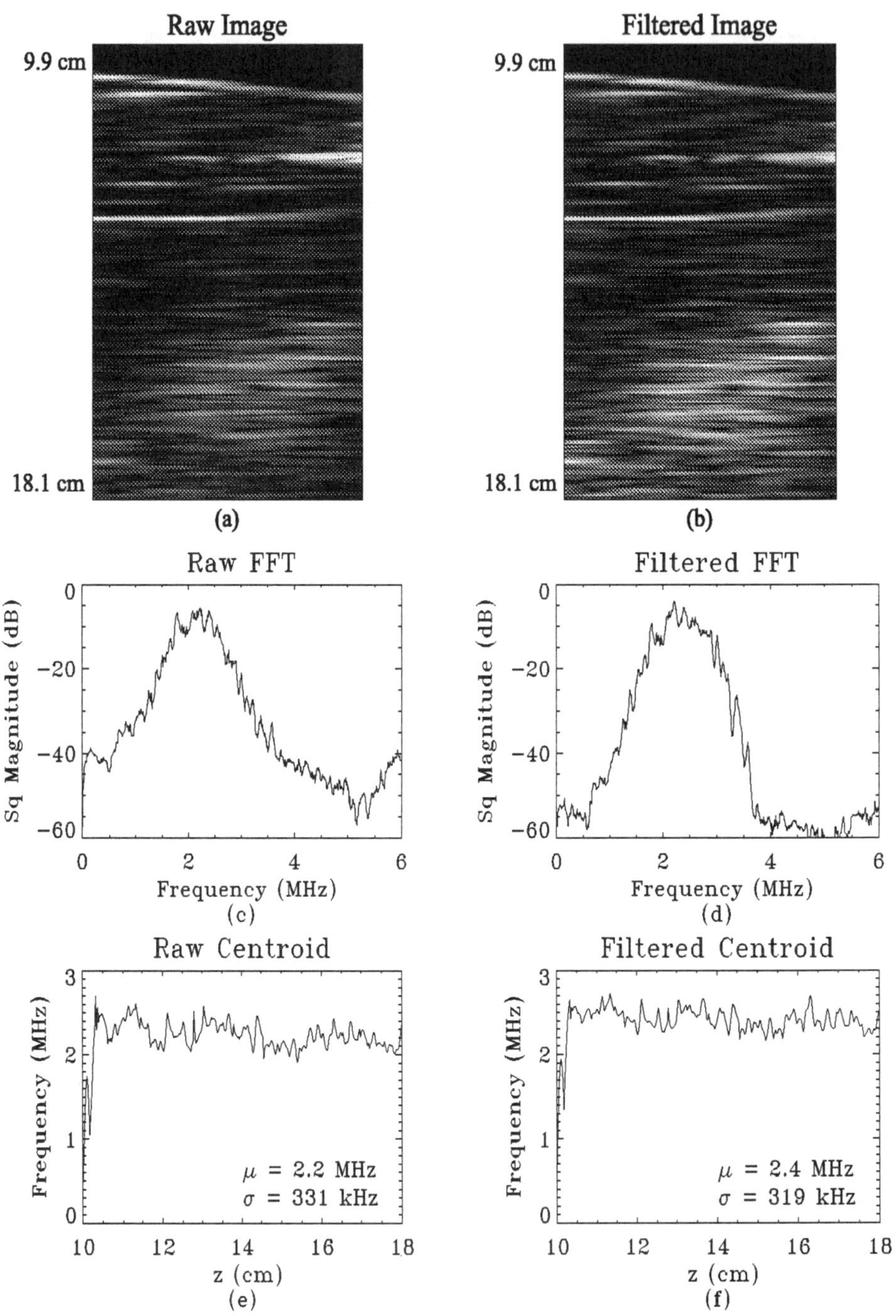

Figure 6.13.   Pig liver on sponge — 2.25 MHz long focus ($2a = 13$ mm).

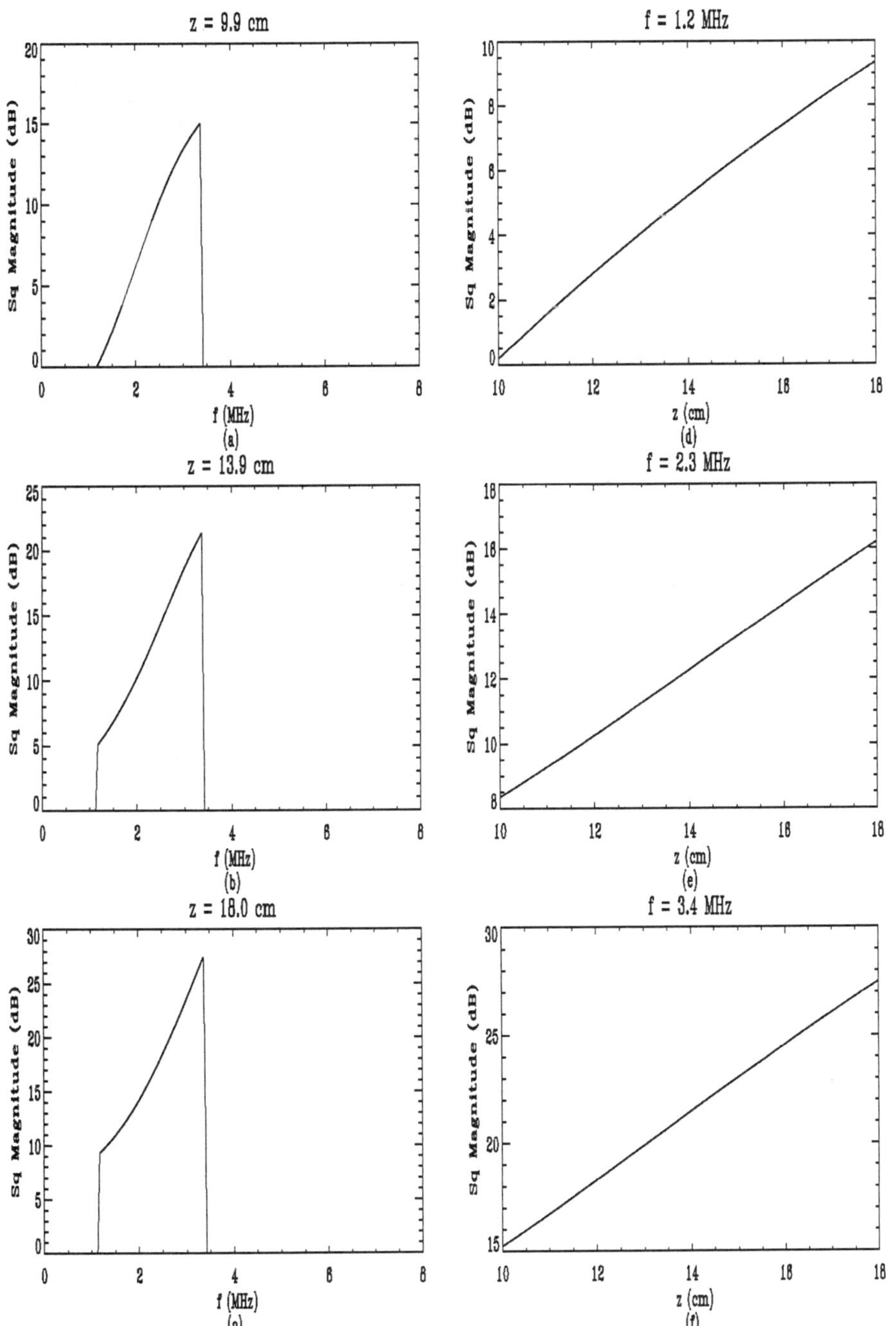

Figure 6.14.   Magnitude response — 2.25 MHz long focus ($2a = 13$ mm).

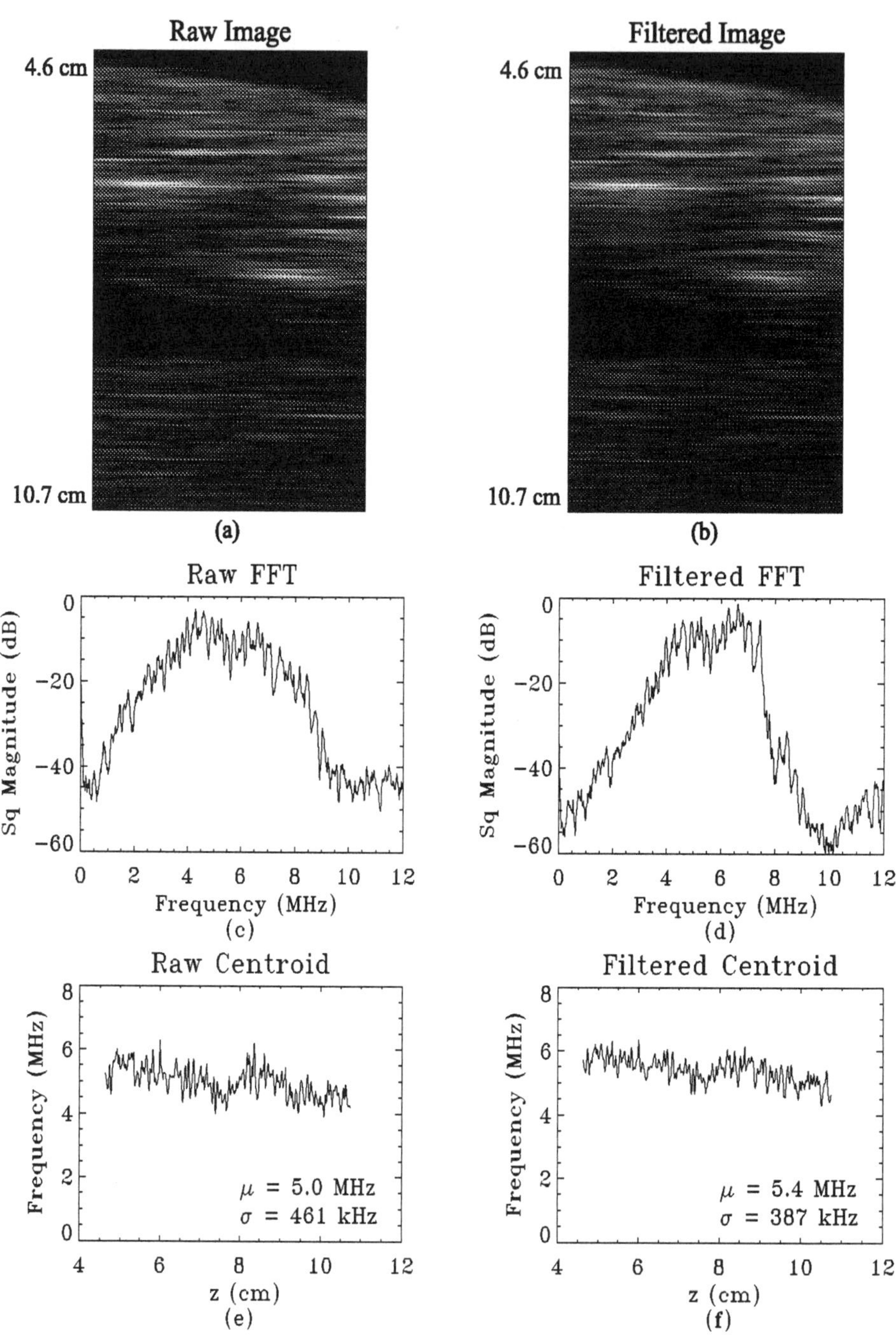

Figure 6.15.   Pig liver on sponge — 5.0 MHz unfocused ($2a = 9.53$ mm).

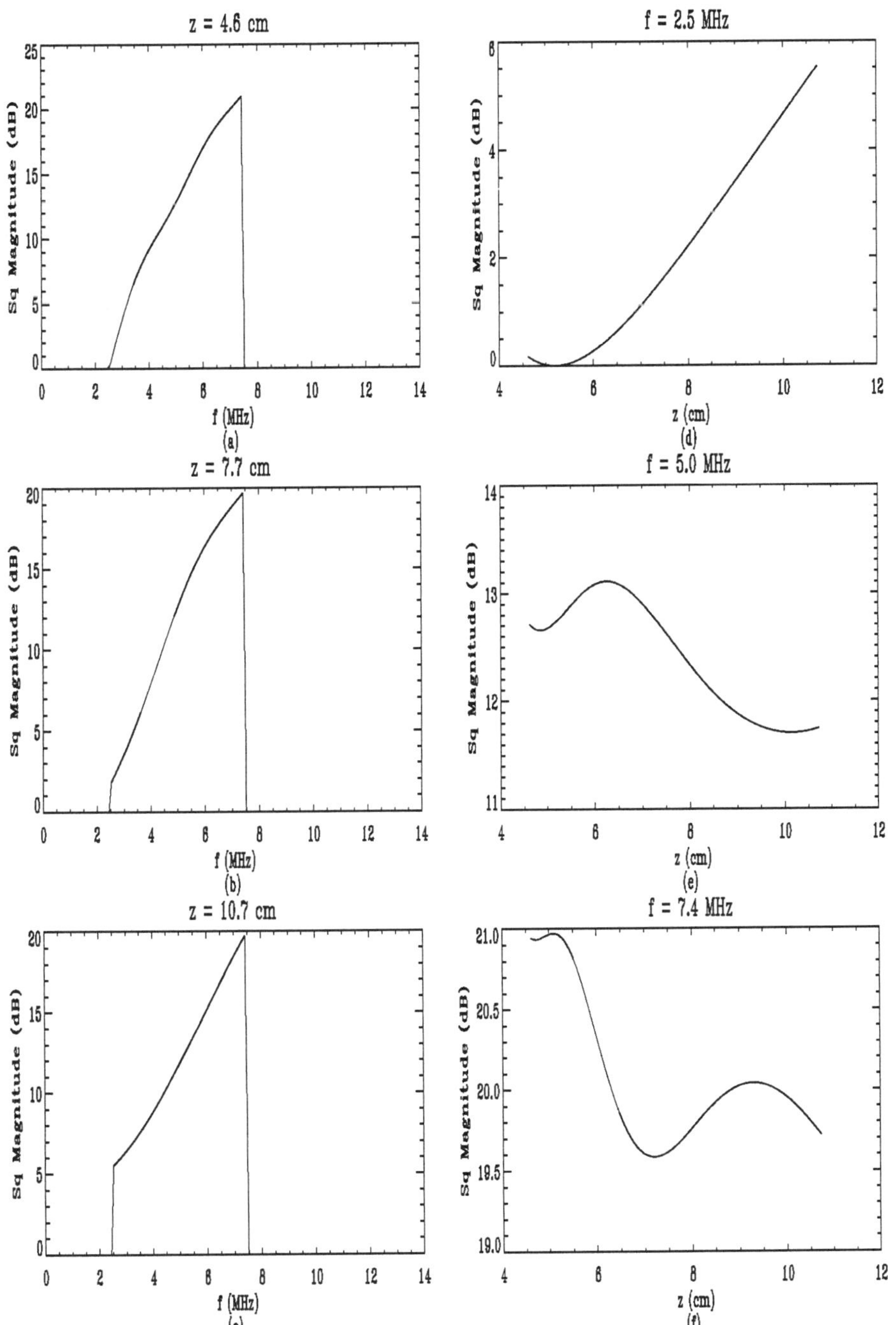

Figure 6.16.   Magnitude response — 5.0 MHz unfocused ($2a = 9.53$ mm).

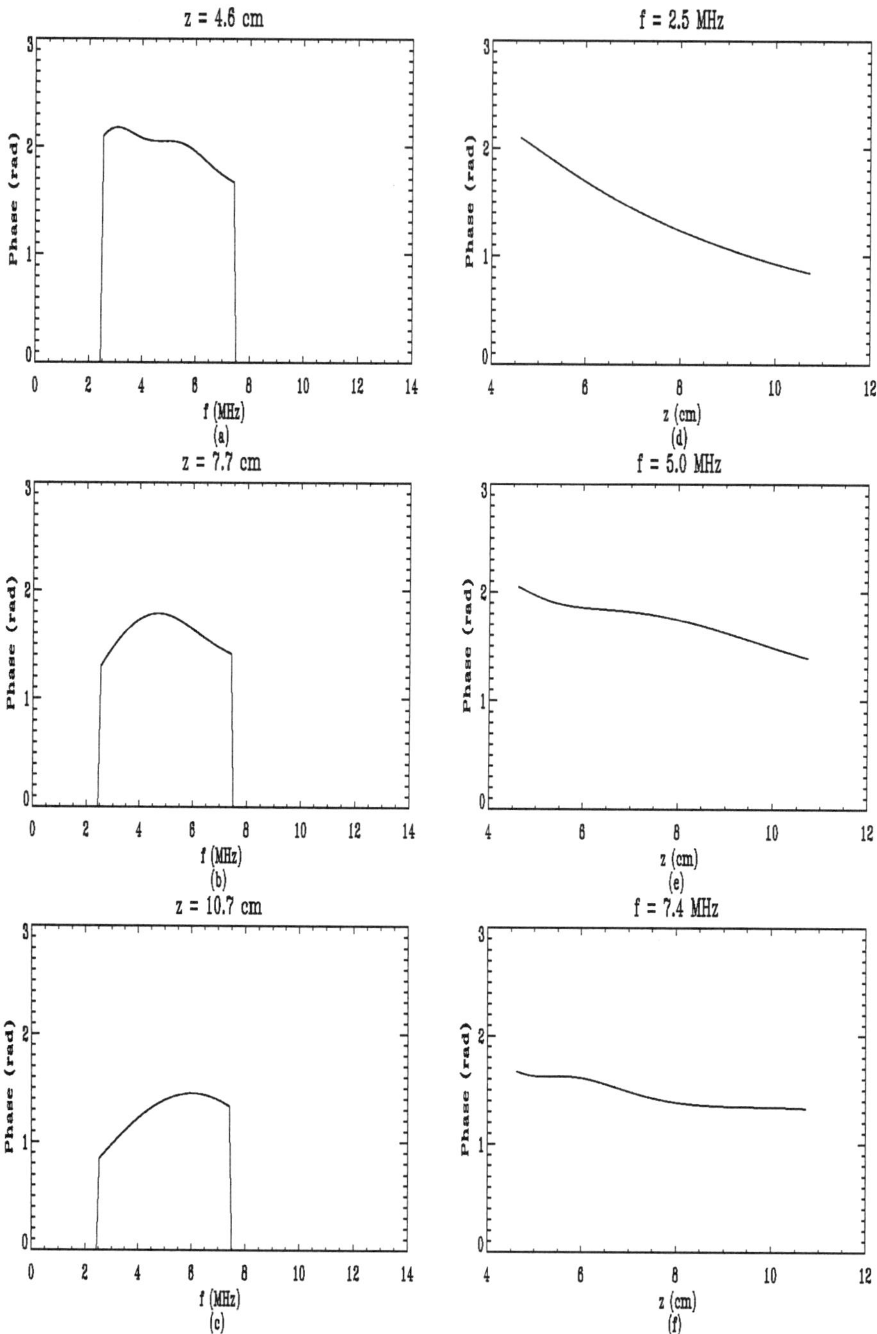

Figure 6.17. Phase response — 5.0 MHz unfocused ($2a = 9.53$ mm).

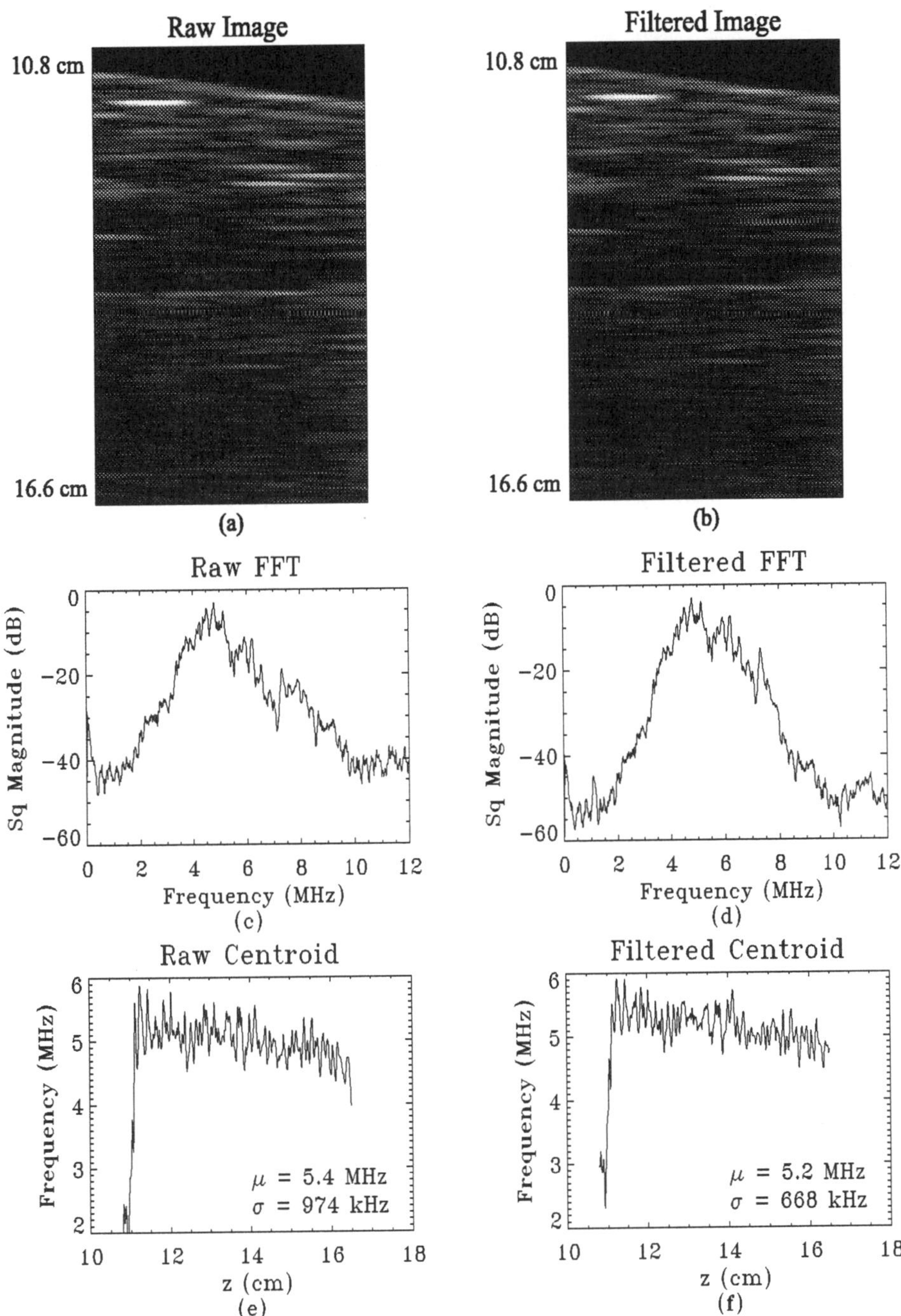

Figure 6.18.   Pig liver on sponge — 5.0 MHz long focus ($2a = 13$ mm).

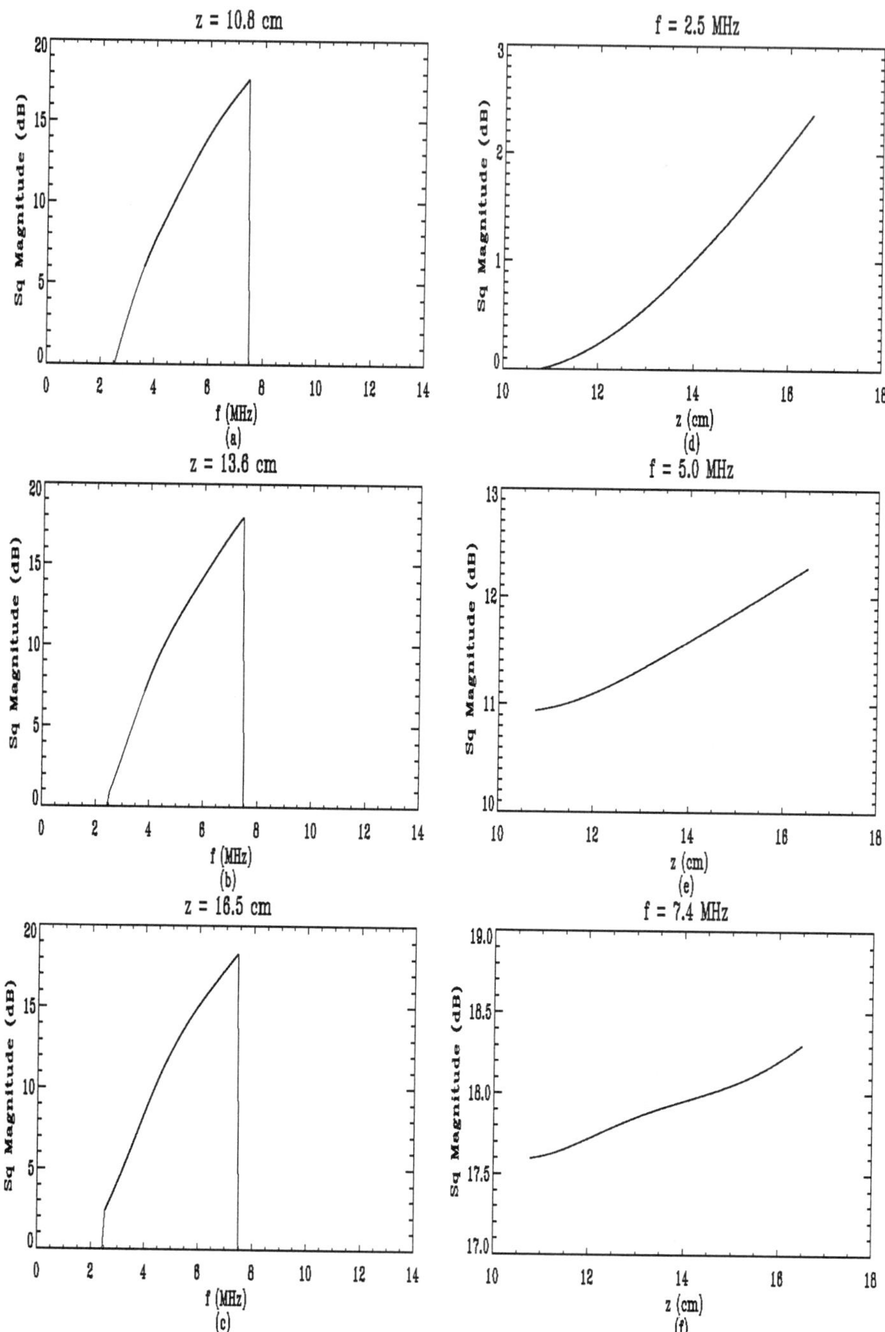

Figure 6.19.   Magnitude response — 5.0 MHz long focus ($2a = 13$ mm).

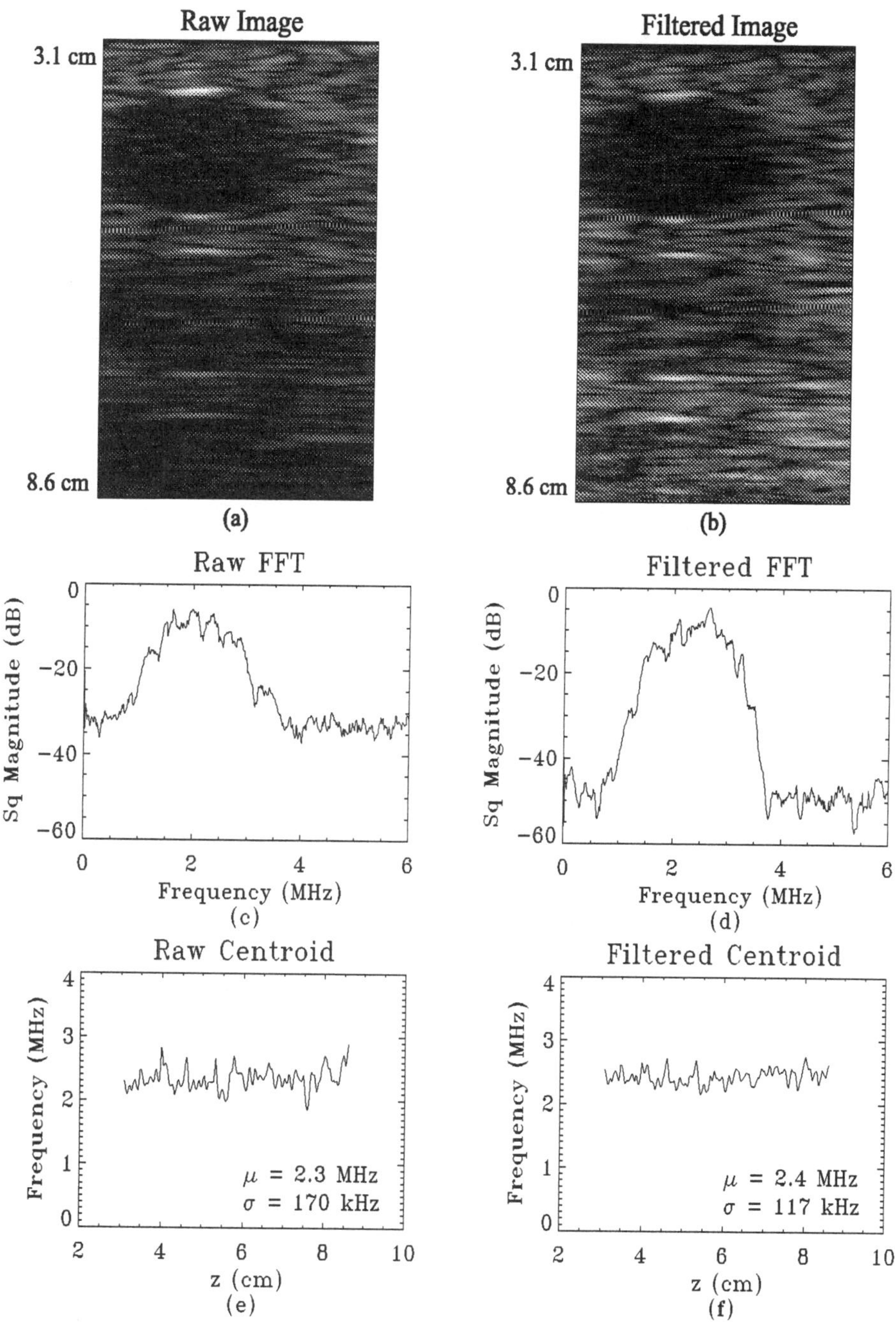

Figure 6.20.   Disk phantom — 2.25 MHz medium focus ($2a = 13$ mm).

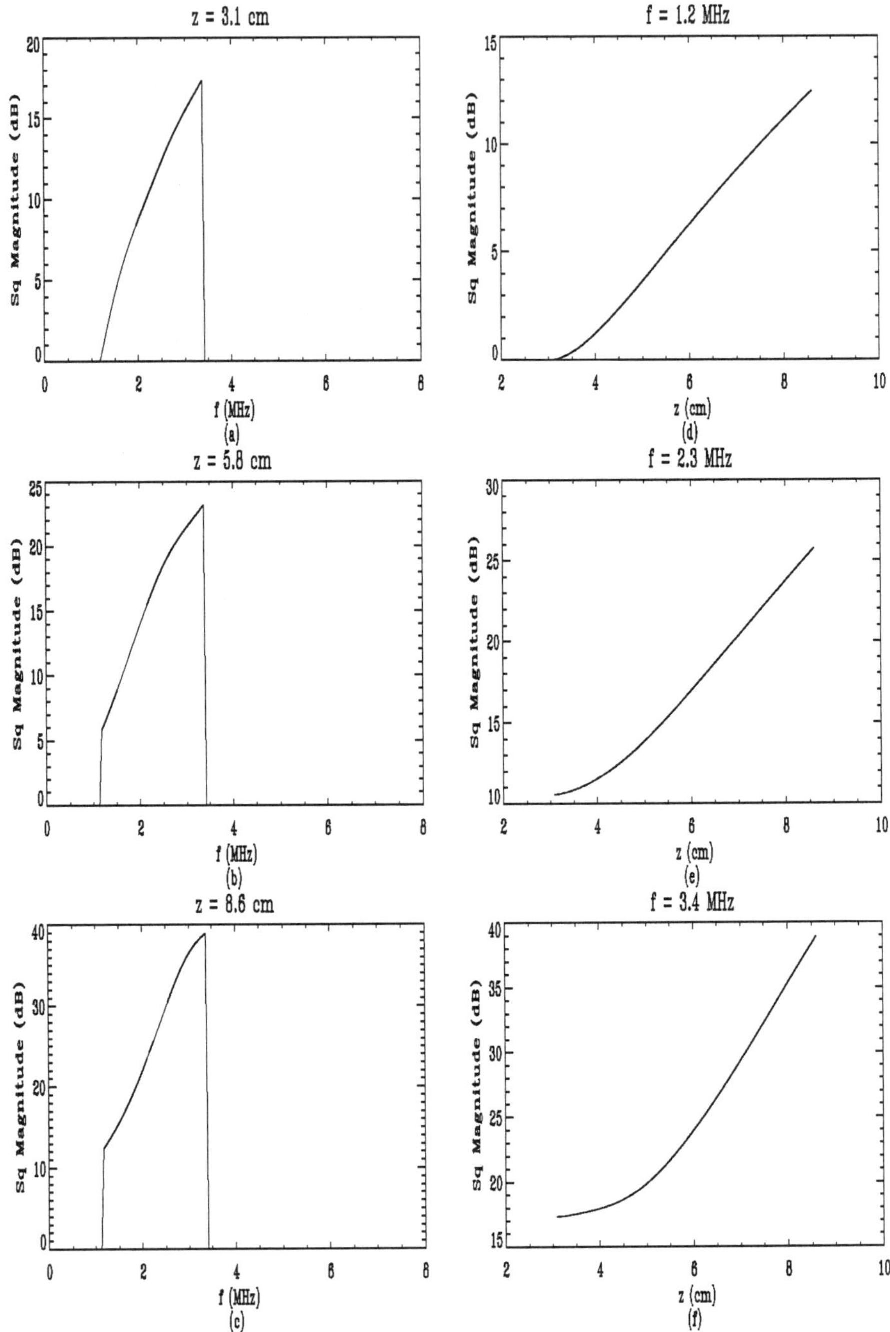

Figure 6.21.   Magnitude response — 2.25 MHz medium focus ($2a = 13$ mm).

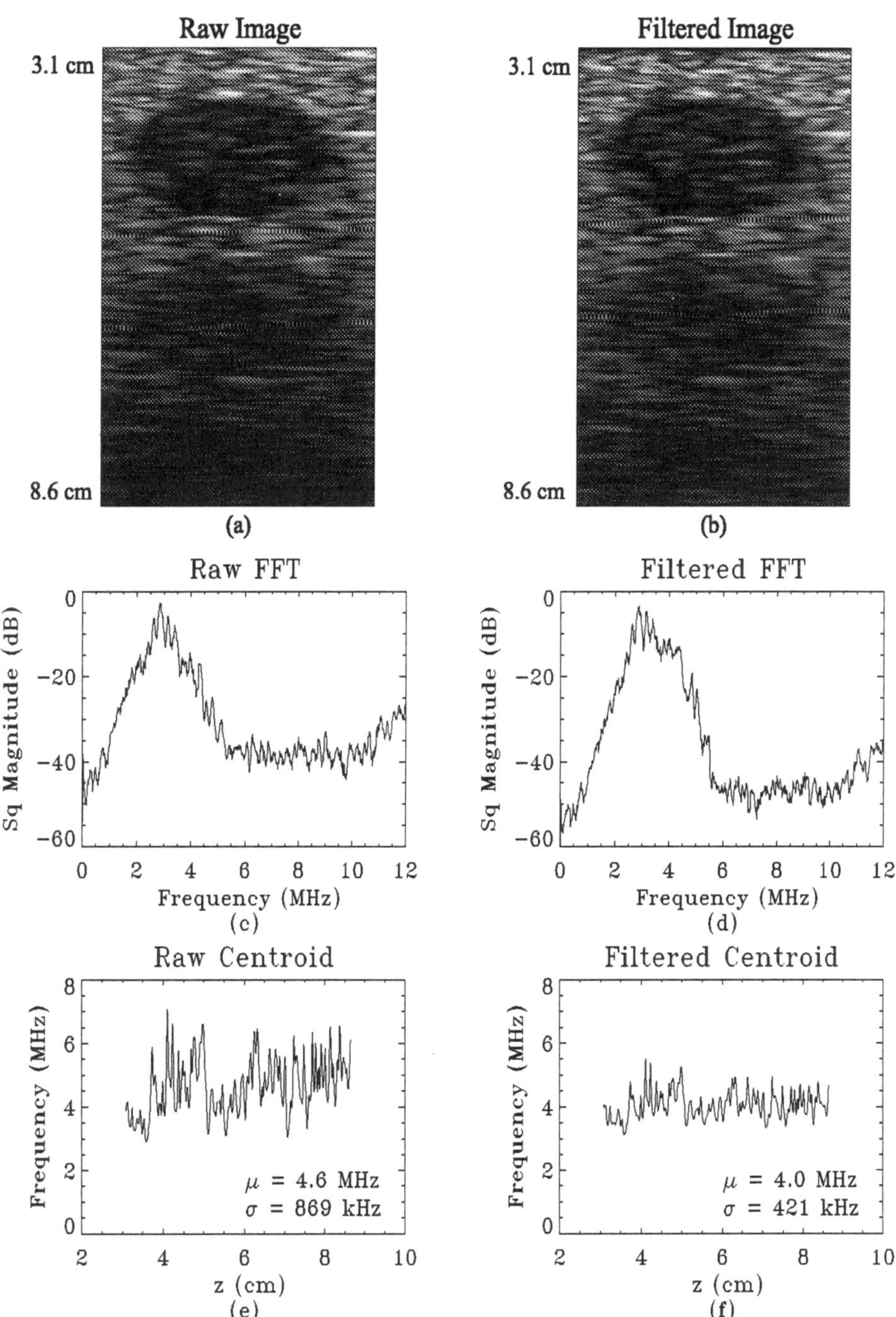

Figure 6.22.   Disk phantom — 3.5 MHz medium focus ($2a = 13$ mm).

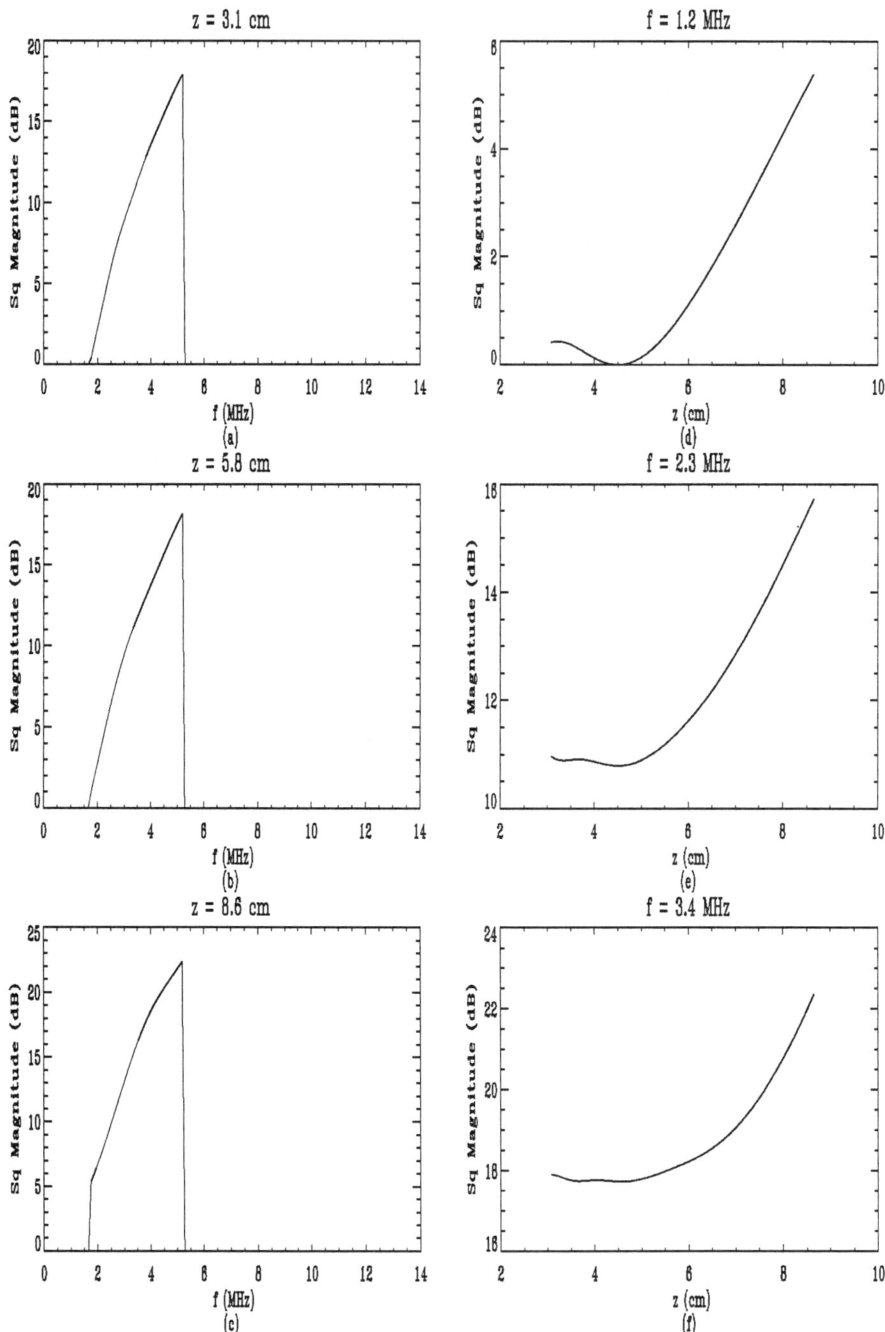

Figure 6.23.   Magnitude response — 3.5 MHz medium focus ($2a = 13$ mm).

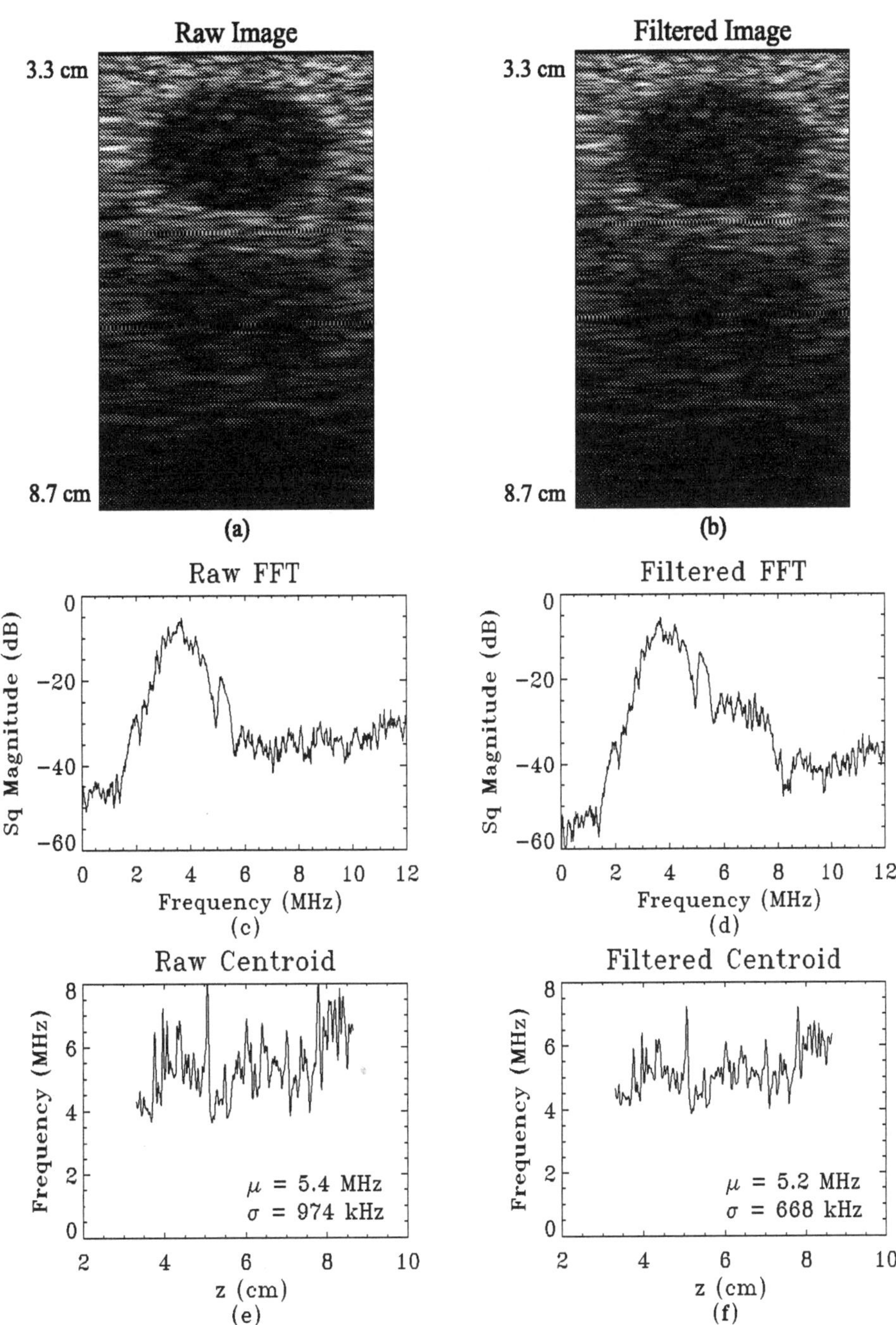

Figure 6.24.   Disk phantom — 5.0 MHz medium focus ($2a = 13$ mm).

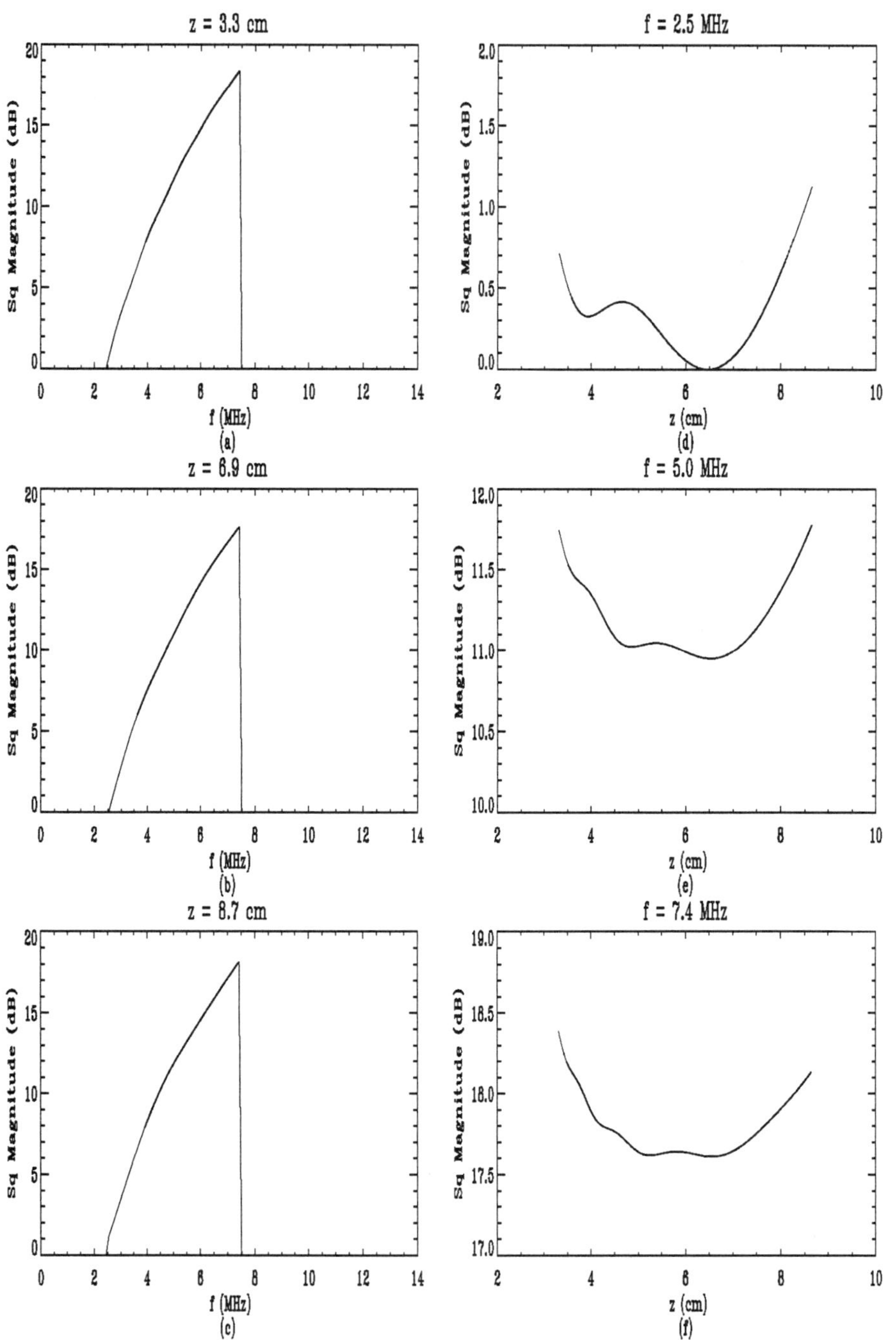

Figure 6.25.   Magnitude response — 5.0 MHz medium focus ($2a = 13$ mm).

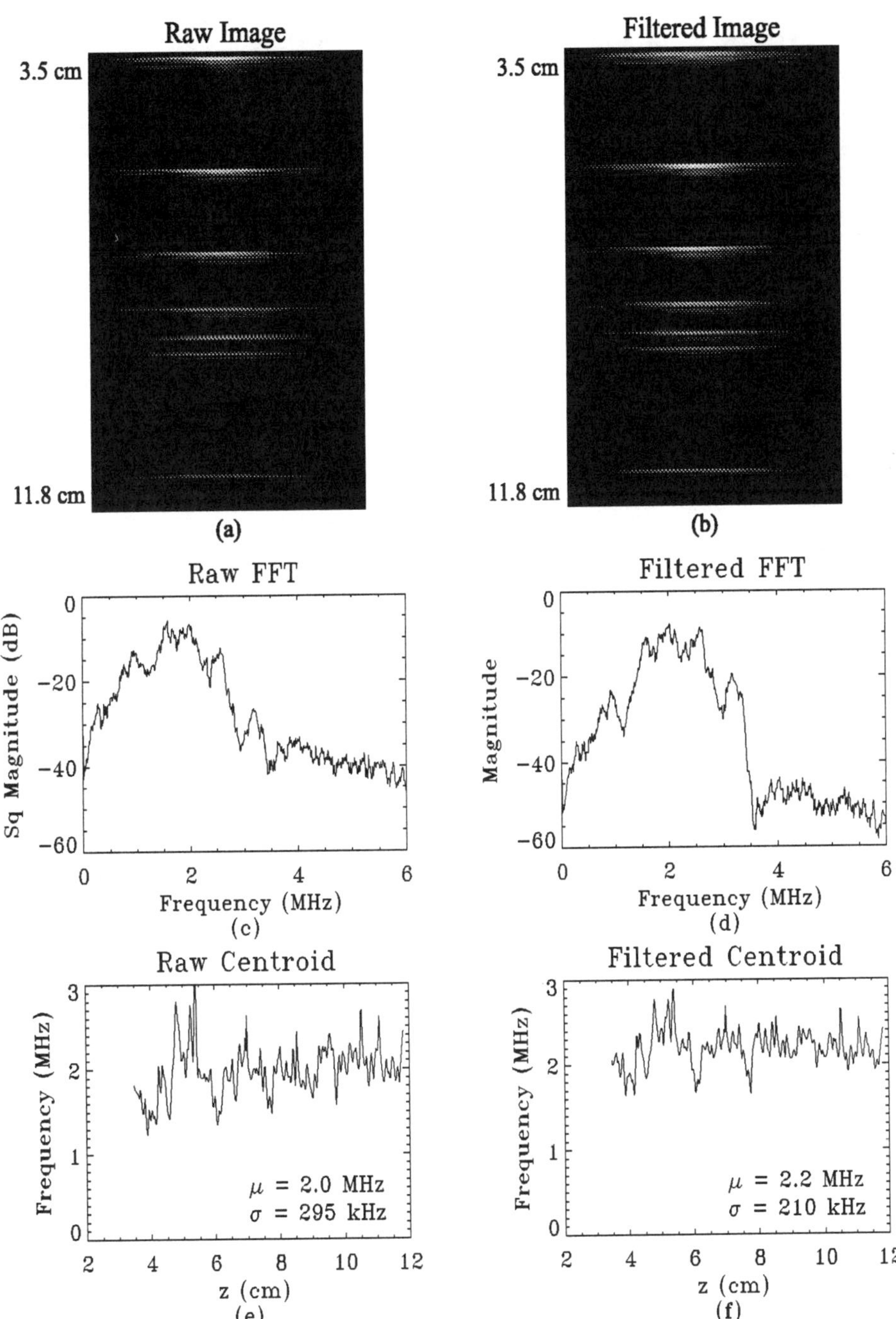

Figure 6.26.   Wire targets — 2.25 MHz unfocused ($2a = 13$ mm).

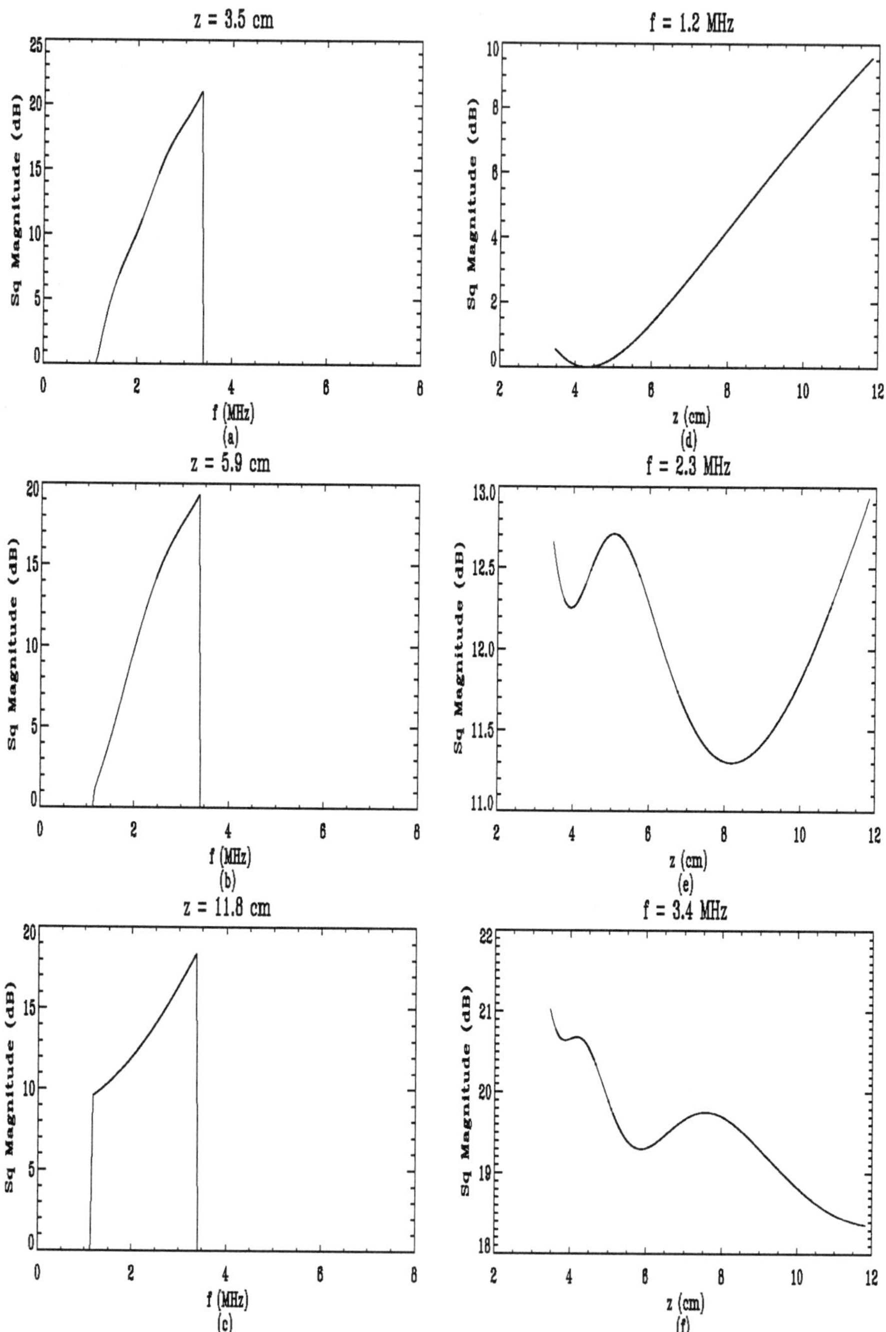

Figure 6.27. Magnitude response — 2.25 MHz unfocused ($2a = 13$ mm).

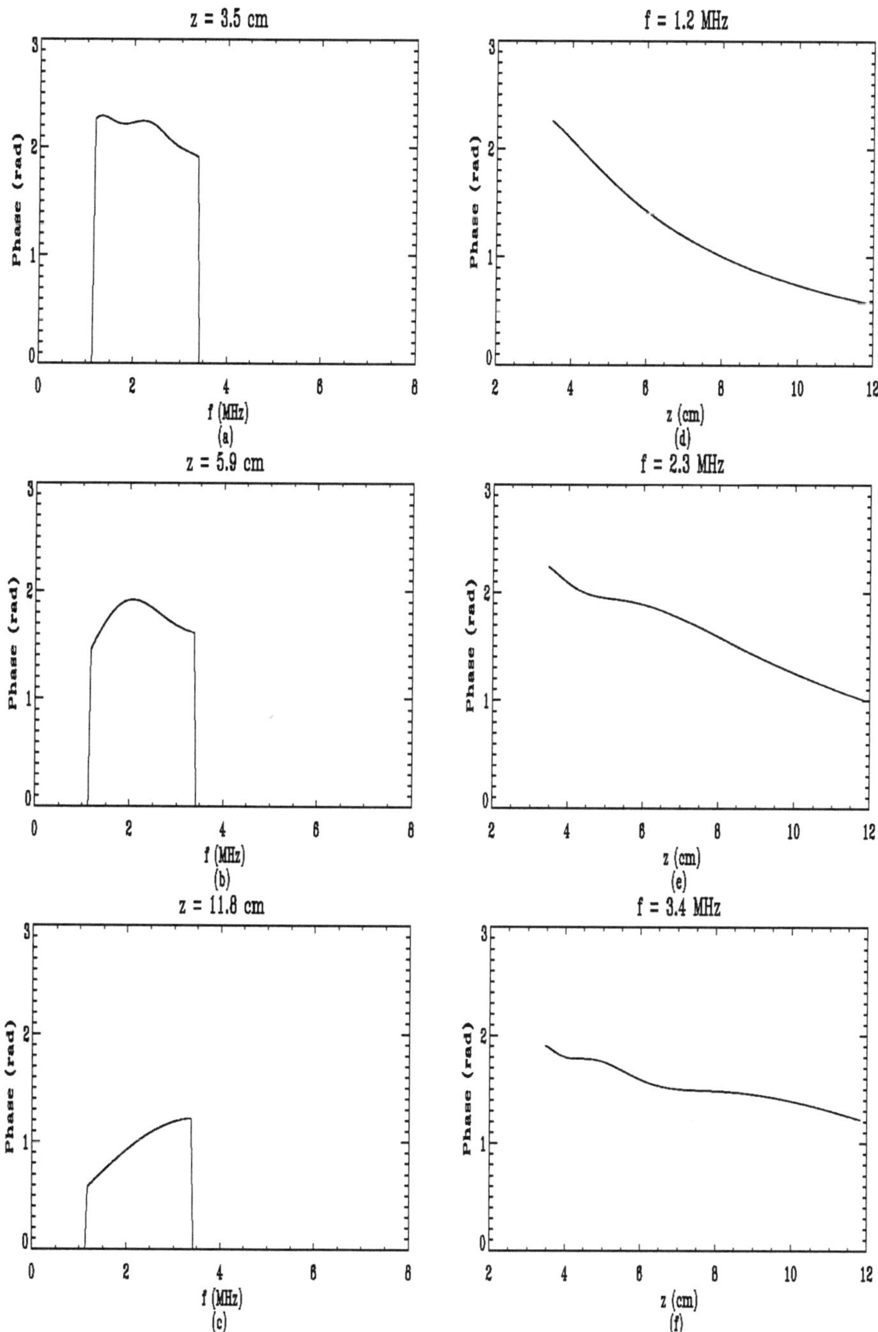

Figure 6.28.   Phase response — 2.25 MHz unfocused ($2a = 13$ mm).

Chapter 7

ANALYTICAL INVESTIGATION

The arccos and Lommel diffraction formulations were connected as an approximate Fourier transform pair in Chapter 3. Closed-form spatially averaged diffraction corrections for one-way and two-way diffraction were presented in Chapter 4 and Chapter 5, and an experimental application of two-way diffraction correction was investigated in the last chapter. In this chapter, we re-examine the theory developed thus far and glean some final results and insights from it.

1. DIFFRACTION AND LINEAR MODELS

In this section, we concern ourselves with indirect experimental verification of the diffraction theory developed in previous chapters. Specifically, we will incorporate spatially averaged diffraction theory in linear models to simulate experiments done by Weight and Hayman [93]. Both one-way and two-way diffraction for a point receiver/scatterer will be analyzed. The one-way case is treated first.

Consider an ideal piston transducer excited by one cycle of a sinusoid of a given frequency and a theoretical point receiver located some coaxial distance z from the transducer. This situation was investigated experimentally by Weight and Hayman in 1978. The researchers investigated the response of a 75μm- radius, wideband receiving element in the beam of an 8-mm radius, wideband unfocused piston transducer. The transducer was excited by a single cycle of a 3-MHz sinusoid. In the framework of our proposed theory, the mathematics of this one-way experiment may be described with a one-way linear model,

$$v_R(z,t) = c_1 \left[\frac{\partial v_T(t)}{\partial t} * \langle h_1(z,t) \rangle_b \right] \tag{7.1}$$

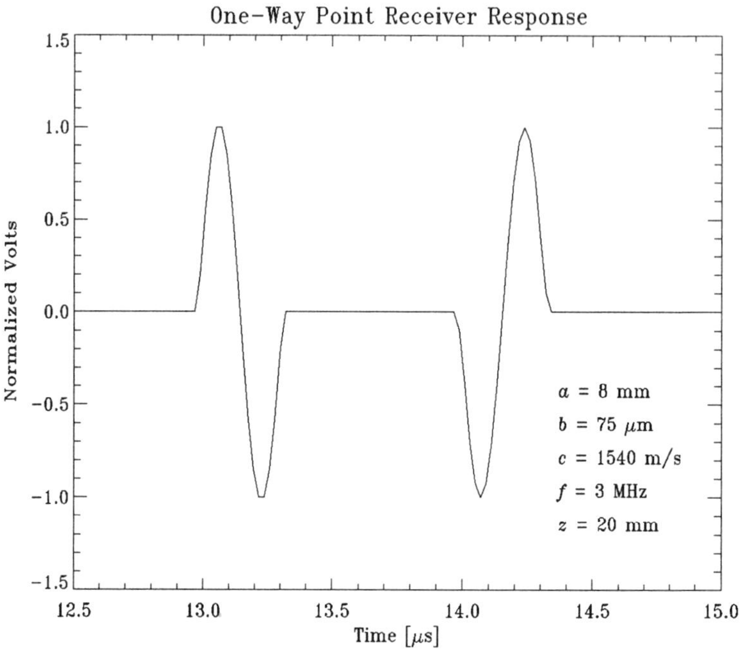

Figure 7.1.   One-way response of an ideal point receiver: On- axis.

where c_1 captures any constants of proportionality, $z = 20$ mm, $b = 75\mu$m, and the remaining terms are the same as in Eq. 5.1. Note the first derivative is used in Eq. 7.1 because we are considering one-way diffraction. Fig. 7.1 shows a plot of the results obtained by applying the one-way model in Eq. 7.1.

The results compare quite favorably to the theoretical and experimental results reported by Weight and Hayman; indeed, the theoretical results are virtually identical. Similar results would have been obtained had we used $\langle \hat{h}_1(z,t)\rangle_b$ instead of $\langle h_1(z,t)\rangle_b$ in Eq. 7.1. However, any derivative of other than first order would give very different results. Furthermore, it is important to emphasize that the results shown in Fig. 7.1 are based on spatially averaged diffraction theory, *not* on point theory.

Attention is now focused on two-way diffraction. Consider an ideal piston transducer excited by one cycle of a given frequency and a theoretical point scatterer located at some coaxial distance z from the transducer. This situation was also investigated experimentally by Weight and Hayman. The researchers investigated the response obtained by insonifying a 0.4-mm radius, on-axis disk with 4-MHz single-cycle exci-

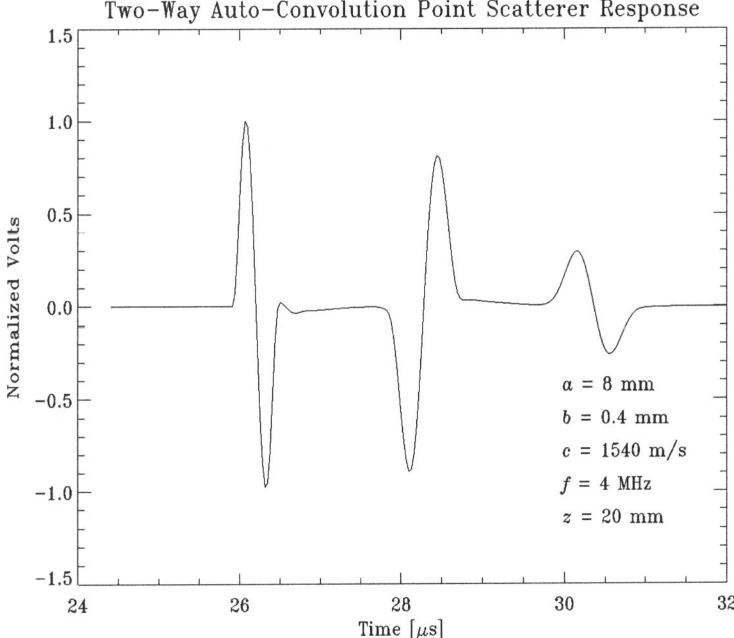

Figure 7.2. Ideal transducer response to on-axis point scatterer: autoconvolution.

tation of a 8-mm radius, unfocused wideband transducer operating in pulse-echo mode.

In the autoconvolution interpretation of reflection imaging, the mathematics of this two-way experiment may be described with a two-way autoconvolution model,

$$v_R(z,t) = c_2 \left[\frac{\partial^2 v_T(t)}{\partial t^2} * \langle \hat{h}_2(z,t) \rangle_b \right], \qquad (7.2)$$

where c_2 captures any constants of proportionality, $z = 20$ mm, $b = 0.4$ mm, $\langle \hat{h}_2(z,t) \rangle_b$ is from Eq. 5.25, and the remainder of the terms are the same as before. It is important to note that Eq. 7.2 is virtually identical to Eq. (14) in Reference [27].

Fig. 7.2 shows that results obtained from Eq. 7.2 compare favorably, in terms of number of pulses, polarity of pulses, and time separation between pulses, to the theoretical and experimental results reported by Weight and Hayman. There is disagreement, however, between the pulse amplitudes we've computed and those reported by Weight and Hayman. This disagreement requires further study. However, it is again important to emphasize that the results shown in the figure are based on a spatially averaged diffraction theory, *not* on point theory.

2. HARMONIC IMAGING AND NON-LINEAR ULTRASOUND

Non-linear theory and techniques developed some forty years ago are finding wide application in modern ultrasonic imaging [18, 39]. Indeed, state-of-the-art ultrasonic equipment relies on some form of these non-linear techniques in an imaging modality referred to in the literature as *harmonic imaging*. From a theoretical standpoint, annular arrays figure prominently in the research on harmonic imaging; thus, both the arccos and Lommel diffraction formulations have application in this new area of research. Only the Lommel formulation is discussed in this section; the interested reader is encouraged to pursue the spatially averaged arccos and Lommel diffraction formulations as they may apply to annular arrays [2, 8, 14].

Harmonic imaging relies on the fact that, in ultrasonic pulse-echo imaging, the insonifying pulse undergoes finite amplitude distortion. This non-linear distortion gives rise to harmonics, and images can be formed using these non-linearly generated higher harmonics.

In an experimental investigation of harmonic imaging, Christopher used a piston transducer that housed two concentric elements. The inner receiving element was a circular disk with a diameter of 1.27 cm; the outer transmitting element was an annulus with inner and outer diameters of 1.27 cm and 1.90 cm, respectively. Transmission was at 2.5 MHz and harmonic reception at 5.0 MHz. Both the transmitting and receiving elements were focused at a depth of 5.08 cm. As part of his experimental investigation, Christopher used a computational modeling protocol to predict various monochromatic beam profiles associated with the transducer just described. The focused Lommel diffraction formulation in Eq. 3.22 can be used to make the same predictions made by Christopher.

In order to do so, we introduce the notation $\widehat{H}_{1,a}(\rho, z, \omega)$ where the subscript a indicates the radius of the piston transducer in question. This notation explicitly specifies the radius of the transducer in Eq. 3.22. Since linear superposition holds, the one-way transmit beam profile $\widehat{H}_T(\rho, z, \omega)$ of the annulus may be written

$$\widehat{H}_T(\rho, z, \omega) = \widehat{H}_{1,b}(\rho, z, \omega) - \widehat{H}_{1,a}(\rho, z, \omega), \qquad (7.3)$$

where a and b are the inner and outer diameters, respectively, of the annulus. Not surprisingly, the one-way receive beam profile $\widehat{H}_R(\rho, z, \omega)$

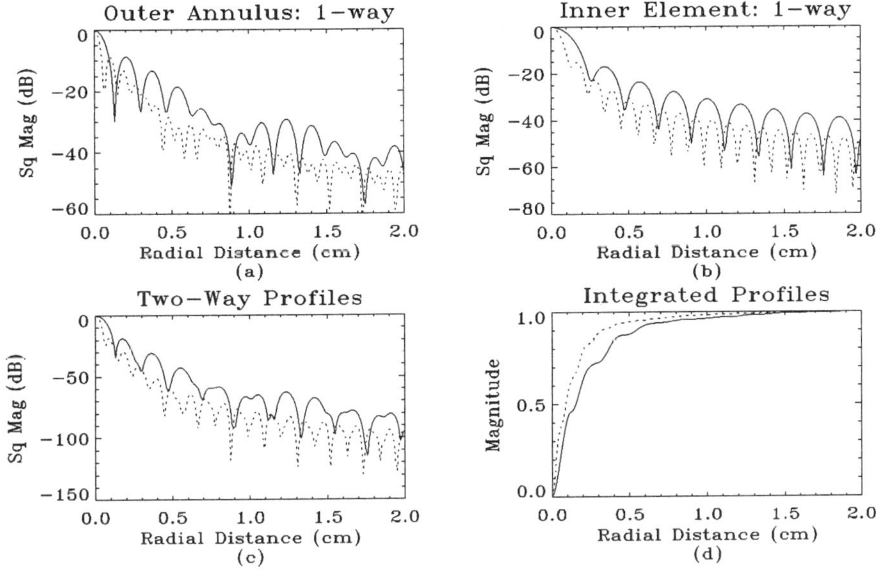

Figure 7.3.   Beam profiles after Christopher [19] for 2.5 MHz fundamental (solid) and 5.0 MHz harmonic (dashed).

of the receiving element may be written

$$\widehat{H}_R(\rho, z, \omega) = \widehat{H}_{1,a}(\rho, z, \omega). \tag{7.4}$$

Eqs. 7.3–7.4 were used to compute the various beam profiles computed by Christopher for a depth of $z = 4.4$ cm; a was set at 1.27 cm and b was set at 1.90 cm. The results are shown in Fig. 7.3. Fig. 7.3(a) shows the one-way 2.5 MHz fundamental and 5.0 MHz harmonic beam profiles for the outer annulus of the transducer. Likewise, Fig. 7.3(b) shows the one-way 2.5 MHz and 5.0 MHz beam profiles for the inner element of the transducer. Shown in Fig. 7.3(c) are the two-way 2.5 MHz fundamental and 5.0 MHz harmonic beam profiles for the outer annulus of the transducer. Following Christopher, the two-way profiles were obtained by simply multiplying the corresponding one-way profiles.

Finally, Fig. 7.3(d) shows the radially integrated two-way profiles. The results shown in Fig. 7.3(a)–(d) are in excellent agreement with those found in Fig. 4 of Christopher's 1998 paper [19]. The reader is referred to Christopher's work for more detailed discussions on the integrated profiles and non-linear ultrasound. These results show that the theory developed in previous chapters can be applied to harmonic imaging.

3.   FOCUSED ONE-WAY RESULTS

Focused one-way results were conspicuous in their absence in Chapter 4. Simply put, spatially averaged one-way results for the focused case were not derived because the algebra involved in solving the required integrals is very messy. Consider

$$\langle \widehat{H}_1(z,\omega) \rangle_b = \frac{1}{\pi b^2} \left[2\pi \int_0^b \widehat{H}_1(\rho,z,\omega)\, \rho\, d\rho \right], \qquad (7.5)$$

where $\widehat{H}_1(\rho,z,\omega)$ is the focused Lommel diffraction formulation. Finding a closed-form solution for this integral is a difficult because the $1/\epsilon$ term in the focused Lommel diffraction formulation introduces a messy asymmetry into the integral.

Suppose, though, that we are content with finding the spatially averaged intensity in the beam of a focused transducer. The intensity of the focused Lommel diffraction formulation is

$$I = \left| \widehat{H}_1(\rho,z,\omega) \right|^2 = \left(\frac{\epsilon}{kz} \right)^2 \left[U_1^2 \left(\frac{ka^2}{\epsilon}, \frac{ka\rho}{z} \right) + U_2^2 \left(\frac{ka^2}{\epsilon}, \frac{ka\rho}{z} \right) \right].$$
$$(7.6)$$

The reader is referred to Eq. 3.22 and the accompanying text for an explanation of terms.

Spatially averaging the velocity-potential intensity in Eq. 7.6 over an arbitrary disk or measurement circle yields

$$\langle I \rangle_b = \frac{1}{\pi b^2} \left\{ 2\pi \left(\frac{\epsilon}{kz} \right)^2 \int_0^b \left[U_1^2 \left(\frac{ka^2}{\epsilon}, \frac{ka\rho}{z} \right) + U_2^2 \left(\frac{ka^2}{\epsilon}, \frac{ka\rho}{z} \right) \right] \rho\, d\rho \right\},$$
$$(7.7)$$

where, as before, the symbol $\langle \rangle$ subscripted with b denotes spatial averaging over a disk of radius b and the angular integration from 0 to 2π has been completed. Eq. 7.7 is based on the assumption that the focused transducer of radius a and the measurement circle of radius b are coaxially located, parallel to one another, and separated by a distance z.

The substitutions $u = ka/\epsilon$, $v = ka\rho/z$, and $v_b = kab/z$ allow Eq. 7.7 to be written more compactly:

$$\langle I \rangle_b = \left(\frac{a}{kb} \right)^2 \left\{ \frac{2}{u^2} \int_0^{v_b} \left[U_1^2(u,v) + U_2^2(u,v) \right] v\, dv \right\}. \qquad (7.8)$$

The braced integral in Eq. 7.8 is the same integral we encountered in Eq. 5.19. As a result, we may once again borrow Wolf's results.

Thus, the integral in Eq. 7.8 may be expressed in terms of the results derived by Wolf for three cases of v_b: $v_b < u$, $v_b = u$, and $v_b > u$. The results are, for $v_b < u$,

$$\langle I \rangle_b = \left(\frac{a}{kb}\right)^2 \left(\left(\frac{v_b}{u}\right)^2 \left[1 + \sum_0^\infty \frac{(-1)^s}{2s+1} \left(\frac{v_b}{u}\right)^{2s} Q_{2s}(v_b)\right] \right.$$
$$\left. - \frac{4}{u}\left[Y_1(u,v_b)\cos\left(\frac{u}{2}+\frac{v_b}{2u}\right) + Y_2(u,v_b)\sin\left(\frac{u}{2}+\frac{v_b}{2u}\right)\right]\right); \quad (7.9)$$

for $v_b = u$,

$$\langle I \rangle_b = \left(\frac{a}{kb}\right)^2 \left(1 - J_0(u)\cos(u) - J_1(u)\sin(u)\right); \quad (7.10)$$

and for $v_b > u$,

$$\langle I \rangle_b = \left(\frac{a}{kb}\right)^2 \left(1 - \sum_0^\infty \frac{(-1)^s}{2s+1}\left(\frac{u}{v_b}\right)^{2s} Q_{2s}(v_b)\right). \quad (7.11)$$

It is important to note that these results reduce to the case of an unfocused transducer in the limit as $A \to \infty$ because $\lim_{A\to\infty} \epsilon = z$.

Eqs. 7.9–7.11 were computed for six values of the parameter b in the case of a medium ($A = 0.5a^2/\lambda = 0.5Z$) focus transducer [53]; the results are shown in Fig. 7.4. The solid-line graphs in Fig. 7.4 show the spatially averaged intensity plotted as a function of $S = z\lambda/a^2$ for each case. The dotted line in each graph is a plot of the axial intensity of the velocity potential (Eq. 3.26) normalized to its maximum value and plotted as a function of S. All results shown in Fig. 7.4 are plotted in decibels (dB).

Fig. 7.4(a) shows the spatially averaged intensity when $b = a/100$. In this case, the measurement circle is essentially a on-axis point, and the results should be similar to those obtained from Eq. 3.26. The graph in Fig 7.4(a) indicates the results are virtually identical. Indeed, that

$$\lim_{b\to 0}\langle I \rangle_b = \left|\widehat{H}_1(0,z,\omega)\right|^2 \quad (7.12)$$

can be proved analytically, and the fairly simple proof is outlined in the following paragraphs.

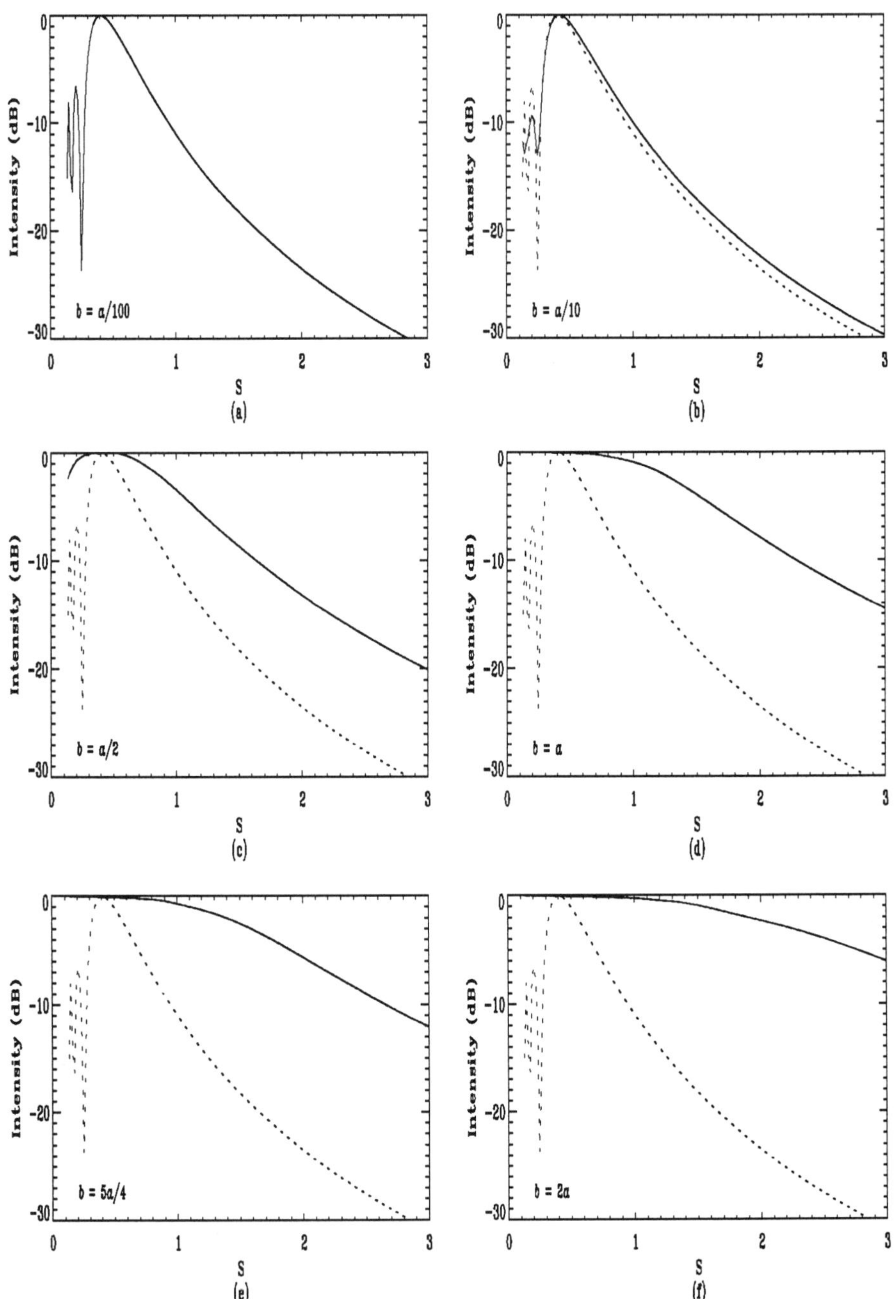

Figure 7.4. Velocity-potential intensity from a medium-focus piston transducer of radius a: spatially averaged over a measurement circle of radius b (solid) and on-axis point intensity (dotted).

Multiplying the right hand side of Eq. 7.9 by $uv_b^2/(uv_b^2)$ and rearranging yields

$$\langle I \rangle_b = \left(\frac{av_b^2}{kbu}\right)^2 \left(\frac{1}{u}\left[1 + \sum_0^\infty \frac{(-1)^s}{2s+1}\left(\frac{v_b}{u}\right)^{2s} Q_{2s}(v_b)\right]\right.$$
$$\left. - \frac{4}{u}\left[\frac{u}{v_b^2}Y_1(u,v_b)\cos\left(\frac{u}{2}+\frac{v_b}{2u}\right) + \frac{u}{v_b^2}Y_2(u,v_b)\sin\left(\frac{u}{2}+\frac{v_b}{2u}\right)\right]\right).$$

$$(7.13)$$

Taking the limit of Eq. 7.13 as b, or equivalently v_b, goes to zero leads to

$$\lim_{b\to 0}\langle I \rangle_b = \frac{\epsilon a^2}{kz^2}\left[\frac{2}{u} - \frac{2}{u}\cos\frac{u}{2}\right],\qquad (7.14)$$

since the weighted summation of $Q_{2s}(v_b)$ goes to unity in the limit and the $Y_1(u,v_b)$ term and $Y_2(u,v_b)$ term go to $0.5\cos(u/2)$ and 0, respectively, in the limit. Finally, various algebraic and trigonometric manipulations lead to the final result:

$$\lim_{b\to 0}\langle I \rangle_b = \left|\frac{2\epsilon}{kz}\sin\left(\frac{ka^2}{4\epsilon}\right)\right|^2 = \left|\widehat{H}_1(0,z,\omega)\right|^2.\qquad (7.15)$$

Fig. 7.4(b) shows the spatially averaged diffraction field when $b = a/10$. The graph indicates that, in the near field at least, the spatially averaged intensity (solid line) does not oscillate as rapidly as does the axial intensity (dotted line). Despite this difference, the far field behavior plotted for each case is quite similar. Note, in particular, that Z, the position of the last axial maximum, occurs at the same value of S for each case. Figs. 7.4(c)-(f) show that the behavior of the spatially averaged intensity rapidly deviates from that of the axial intensity as the radius of the measurement circle b is increased.

The results shown in Fig. 7.4 lead to two conclusions. First, spatial averaging smoothes out velocity-potential oscillations in the near field and decreases the rate of monotonic fall-off in the far field. Second, the behavior of spatially averaged velocity-potential intensity is very similar to that of the on-axis velocity-potential intensity for values of b an order of magnitude greater than $b = a/100$. These observations hold for all degrees of focusing.

Fig. 7.4, just discussed, compares the spatially averaged intensity in the beam of a medium focus transducer with the axial intensity of the same transducer. The plots in Fig. 7.5 illustrate how the spatially averaged intensity (Eqs. 7.9–7.11) varies as the focusing and the radius of

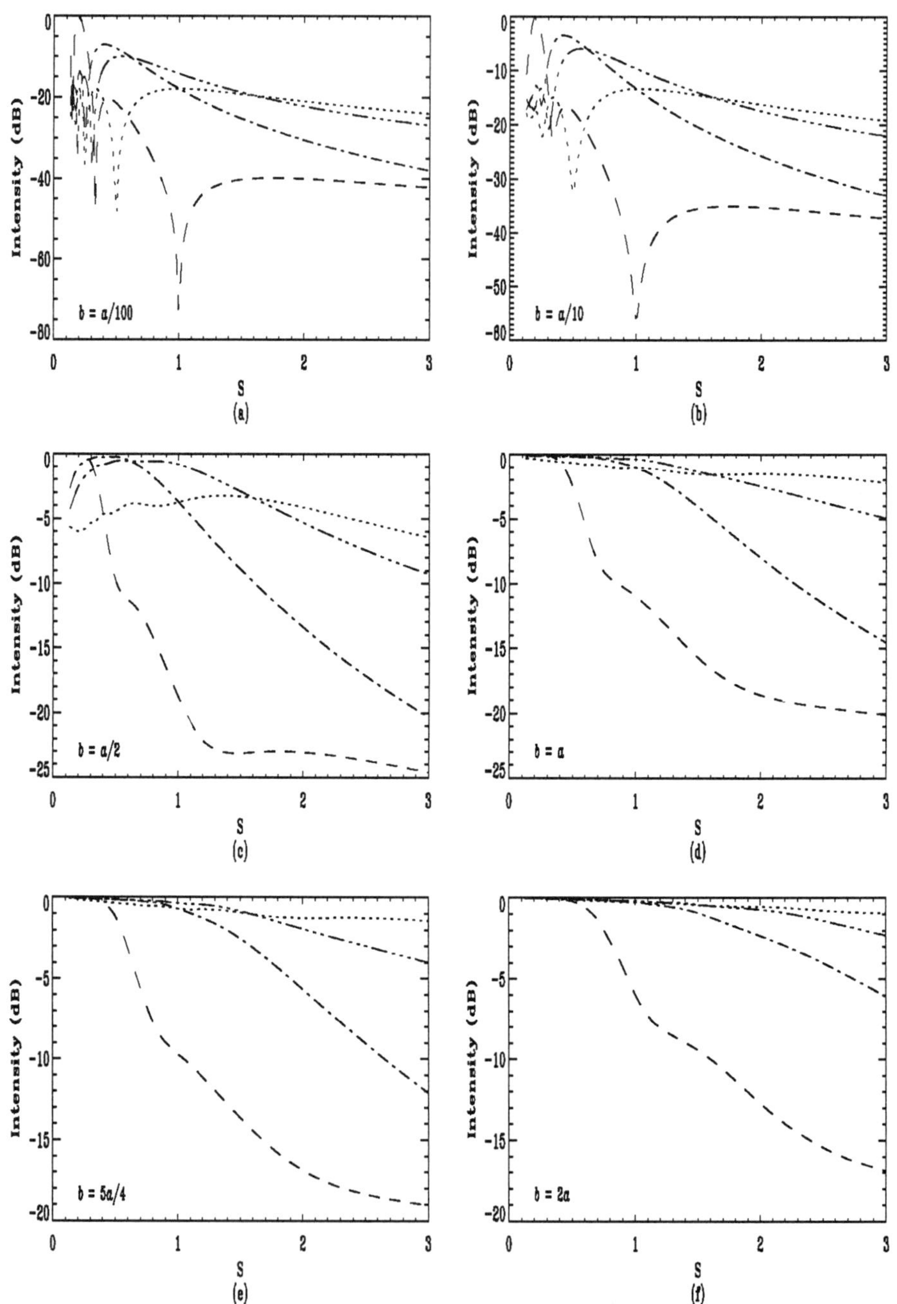

Figure 7.5. Velocity-potential intensity from a piston transducer of radius a as a function of focusing and of measurement radius b: strong focus (dashed), medium focus (dash dot), weak focus (dash double dot), and unfocused (dotted).

the measurement circle b are varied. The value of b is indicated in each graph, and the linestyle corresponds to the degree of focusing: strong (dashed), medium (dash dot), weak (dash double dot), and unfocused (dotted). The following values of A were used: strong ($A = 0.2a^2/\lambda$), medium ($A = 0.5a^2/\lambda$), weak ($A = 0.8a^2/\lambda$), and unfocused ($\epsilon = z$). Recall that strong focusing is also known short focusing, while weak focusing is also known as long focusing.

The results are plotted in dB and are normalized to the maximum value of the strongly focused transducer for each value of b. Not surprisingly, the results illustrated in Fig. 7.4 are consistent with an observation just discussed. Specifically, spatial averaging smoothes out velocity-potential oscillations in the near field, decreases the rate of monotonic fall-off in the far field, and eventually averages the near field out as b is increased. This observation is true for all degrees of focusing.

4. COHERENT VS. INCOHERENT AVERAGING

The previous discussion raises the question: what is the difference between the intensity of the spatially averaged velocity potential and the spatially averaged intensity of the velocity potential? Mathematically posed, the question is how do

$$\left| \langle \widehat{H}_1(z,\omega) \rangle_b \right|^2 = \left| \frac{1}{\pi b^2} \left[2\pi \int_0^b \widehat{H}_1(\rho,z,\omega)\, \rho\, d\rho \right] \right|^2, \qquad (7.16)$$

and

$$\langle \left| \widehat{H}_1(z,\omega) \right|^2 \rangle_b = \frac{1}{\pi b^2} \left[2\pi \int_0^b \left| \widehat{H}_1(\rho,z,\omega) \right|^2 \rho\, d\rho \right] \qquad (7.17)$$

compare? The answer is of theoretical interest and may be of practical interest to those trying to develop devices sensitive to pressure intensity.

An obvious difference is the fact that Eq. 7.16 is a coherent average which includes phase information in the integrand, while Eq. 7.17 is an incoherent average which does not. Thus, we immediately suspect that closed-form solutions to Eq. 7.16 and Eq. 7.17 will not be equal. Indeed, Eqs. 7.9–7.11 and the squared magnitude of Eq. 4.18, which are closed-form solutions to Eq. 7.17 and Eq. 7.16, respectively, cannot be made equal. A curious exception to this statement occurs in the limit as b approaches 0; in this case, the squared magnitude of the coherent average is equivalent to the incoherent average. In other words, the intensity of the spatially averaged velocity potential and the spatially averaged intensity of the velocity potential are identical for an on-axis point receiver. Eq. 7.15 confirms this assertion.

More insight into the nature of coherent and incoherent averaging can be had by examining Fig. 7.6. Fig. 7.6 compares coherent averaging and incoherent averaging in the beam of an *unfocused* piston transducer. Specifically, results obtained Eq. 4.18 and Eqs. 7.9–7.11 are plotted in the figure as a function of $S = z\lambda/a^2$ for six different values of the parameter b with $2a = 13$ mm, $f_c = 2.25$ MHz, and $c = 1540$ m/s. The solid-line graphs illustrate the intensity of the spatially averaged velocity potential, while the dashed-line graphs illustrate the spatially averaged intensity of the velocity potential. All results shown in Fig. 7.6 are squared magnitude results plotted in decibels (dB).

The graphs in Fig. 7.6 lead to five observations, many of which we have already noted in the previous section and in Chapter 4. First, Fig. 7.6(a) graphically corroborates the assertion that the incoherent average and the squared magnitude of the coherent average are identical in the limit as b approaches 0. Second, Figs. 7.6(b)-(c) show that coherent and incoherent averaging are virtually identical, at least in a squared-magnitude sense, for surprisingly large values of the parameter b. Indeed, Figs. 7.6(d)-(f) show that results obtained from coherent and incoherent averaging are within a dB of one another for b approximately equal to a. Third, both coherent and incoherent averaging smooth out oscillations in the near field and decrease the rate of monotonic fall-off in the far field.

The fourth observation requires more explanation. Consider Fig. 7.6(a) and the value of the curve plotted at $S = 1$. Recall that we have agreed to designate the position of the last axial maximum Z. The remainder of the graphs show clearly that Z shifts to the right as b is increased and that the last axial maximum itself is, in a sense, averaged out as b approaches a. This observation holds for both coherent and incoherent averaging.

The final observation concerns Fig. 7.6(a) and Fig. 7.6(e) where we draw the reader's attention to the fact that the value of the curve at $S = 1$ is a local maximum in the former but a local minimum in the latter. Again, this observation holds for both coherent and incoherent averaging. Overall, we conclude that coherent and incoherent averaging in the beam of an *unfocused* piston transducer produce similar squared-magnitude results for $b < a$, the agreement being better with $b \ll a$.

Although these observations are based on *unfocused* results, the observations should also hold for focused transducers. Indeed, we have seen in the previous section that incoherent averaging in the beam of a *focused* transducer smoothes out oscillations in the near field and decreases the rate of monotonic fall-off in the far field. Also, Chen and co-workers have observed that, for a focused transducer operating at fixed frequency, "the

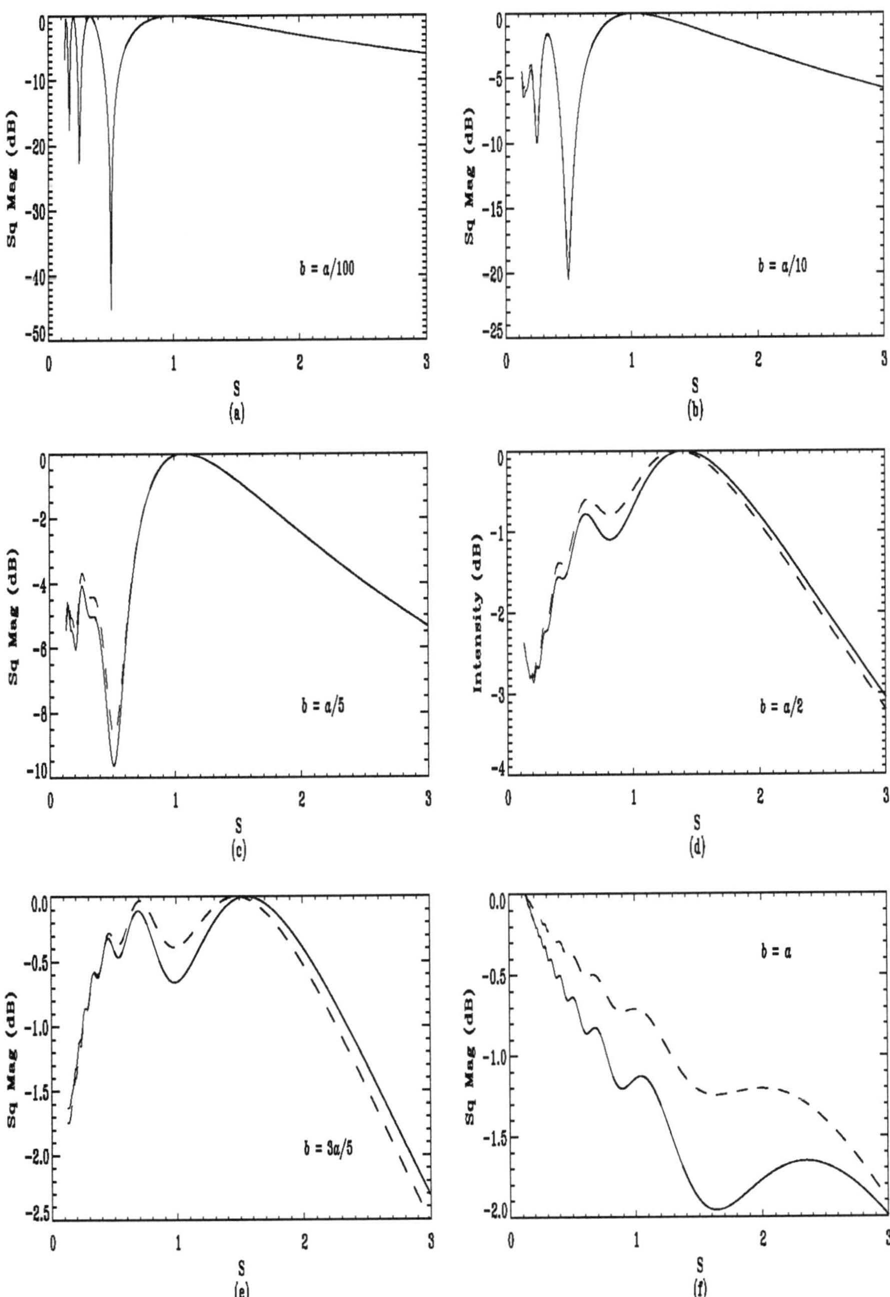

Figure 7.6.   Coherent (solid) vs. incoherent (dotted) spatial averaging as a function of the parameter b.

maximum in [acoustic coupling] occurs farther away from the transducer surface than the on-axis pressure maximum, which is always before the geometrical focus [16]." In other words, the last axial maximum for a focused piston transducer is shifted to the right with spatial averaging.

5.   MIRROR-IMAGE VS. AUTOCONVOLUTION DIFFRACTION

We have already argued in Chapter 5 that results derived by many researchers are ultimately based on or result in an autoconvolution interpretation of two-way diffraction. However, we have also noted repeatedly in previous chapters that certain one-way results can be used to describe reflection from a flat plate by invoking the mirror-image interpretation of two-way diffraction. Indeed, Lizzi and co-workers state that the calibration spectrum from a flat plate can be computed using an autoconvolution integral [59, Eq. (16)] or by applying the method of images—in other words, the mirror-image interpretation of two-way diffraction. Thus, we are forced to ask how the two interpretations compare.

If the two interpretations are the same, then we expect analytic results obtained from the two interpretations to be very similar in both form and interpretation. For example, we would expect Eq. 4.11, with z set to $2z$, to be very similar in form and interpretation to the received calibration pressure P_c derived by Lizzi and co-workers [59, Eq. (17)]. However, such is not the case. Indeed, their result is more similar in form and interpretation to Eq. 4.23, the infinite receiver result derived in Chapter 4, and to Eq. 5.24, the infinite plate result derived in Chapter 5. Oddly, Eq. 4.11 with z set to $2z$ appears to be similar in form and interpretation, at least in a magnitude sense, to Eq. 5.21.

Additionally, the two interpretations are different in terms of applicability. On the one hand, the autoconvolution interpretation holds strictly for a point scatterer. However, we took liberties with the autoconvolution interpretation and extended it by invoking the concept of an effective backscattering cross-section. As a result, we obtained physically meaningful and intuitively appealing results for reflecting disks of any size. On the other hand, the mirror-image interpretation holds only for reflection from an infinite flat plate.

Given the results and discussion presented above and throughout this monograph, we are ultimately forced to answer that the two interpretations are quite different. This answer, in turn, raises the question of which interpretation is correct. The only reasonable answer is that both yield physically meaningful and intuitively appealing results in those instances where they are applicable.

6.   CHAPTER SUMMARY

This chapter developed more results and provided deeper insight into the problem of scalar diffraction from circular aperture. Specifically, linear and non-linear ultrasound were briefly discussed, a focused one-way result was derived, coherent and incoherent averaging were compared, and the mirror-image and autoconvolution interpretations of two-way diffraction were discussed.

Chapter 8

RECOMMENDATIONS FOR FURTHER RESEARCH

This monograph proposed a theory of spatially averaged diffraction correction for ultrasonic piston transducers operating in pulsed mode. A good portion of the theory has been cast in closed-form and verified, both analytically and numerically. Indeed, we even provided a few cases of indirect experimental validation of the theory. Additionally, autoconvolution diffraction corrections were applied experimentally. Despite this, much more work needs to be done.

1. GENERAL

This section points out three general aspects of the theory that demand further research. First and most importantly, large-scale experimentation designed to test the practical limits and clinical applicability of the entire theory is imperative. Second, apodization of the source, an important consideration in practical situations, was not considered; an excellent starting point to remedy this situation is a 1949 paper by Hopkins [43]. The final point concerns computation of the various infinite summations introduced in Chapter 1. In this work, all results were obtained by programming the infinite summations explicitly in truncated do-loops. This brute-force style of programming may be avoided and processing time reduced if recursion relations are used. At any rate, recursion relations for most of the equations in Chapter 1 can be found in the references [38, 90, 97, 98].

2. FOURIER EQUIVALENCE

The Fourier equivalence of the arccos and Lommel diffraction formulations as an approximate Fourier transform pair was rigorously demonstrated for both unfocused and focused piston transducers. Nonetheless, a more detailed study of focused piston transducers is required. Eq. 3.22 gives the Lommel diffraction formulation for a focused piston transducer, and the arccos diffraction formulation for a focused piston transducer can be found in Reference [2]. The study should investigate various degrees of focusing [69] and explain why and where the arccos and Lommel diffraction formulations for focused piston transducers agree and disagree.

3. SPATIALLY AVERAGED ONE-WAY DIFFRACTION

Chapter 4 presented a detailed theoretical treatment of spatially averaged one-way diffraction for unfocused piston transducers. An experimental study of finite receivers is needed to validate the theory; the experiments could be modeled after those in described in Reference [93]. From a theoretical and historical perspective, a study comparing spatially averaged one-way diffraction effects obtained from Eq. 4.11 and previously tabulated values [7, 49] might make a useful undergraduate project.

Spatially averaged incoherent results for focused piston transducers were derived in Chapter 7. However, no coherent results were derived for the focused case. This gaping hole needs to be filled. Doing so, however, will be difficult. Basic probability reveals that a completely general closed-form solution for spatially averaged one-way diffraction would have to cover 36 combinations including the following variations:

- focused or unfocused transmitter: 2 choices,
- focused or unfocused receiver: 2 choices,
- transmitter area $<$, $=$, and $>$ receiver area: 3 choices,
- transmitter focal length $<$, $=$, and $>$ receiver focal length: 3 choices.

Add to this the fact that closed-form results would have to be derived in the time and frequency domains. The strictly unfocused results in Eq. 4.5 and Eq. 4.18 and the focused results derived by Chen, *et al.* [16] would ease the burden somewhat, but clearly, this is a long-term project. Here again, experiments modeled after those in Reference [93] would serve to test the theory.

4. SPATIALLY AVERAGED AUTOCONVOLUTION DIFFRACTION

Chapter 5 presented an *ad hoc* derivation of spatially averaged autoconvolution diffraction effects for both focused and unfocused piston transducers. The approximate Fourier equivalence of the arccos and Lommel diffraction formulations in conjunction with the assumption of minimum phase allowed the derivation of Eq. 5.25. To the extent that Eq. 5.18 is valid, Eq. 5.25 is also valid for both focused and unfocused transducers. Eq. 5.25 was analytically verified for the unfocused case in terms of magnitude and phase and for the focused case in terms of magnitude only.

Thus, a study of focused piston transducers is required. The arccos diffraction formulation for a focused piston transducer can be found in Reference [2], and it would have to be numerically integrated. Eq. 5.25 could be used as is, as long as the time reversal for $z > A$ is observed.

The study should investigate long, medium, and short focusing [69] and explain where and why the arccos and Lommel diffraction formulations for spatially averaged autoconvolution diffraction agree and disagree. Of course, an experimental study of reflection by finite planar disk reflectors is needed to validate the theory; the experiments could be modeled, again, after those done by Weight and Hayman [93].

The derivation of spatially averaged autoconvolution diffraction is lacking in three respects. First, a closed-form time-domain expression for spatially averaged autoconvolution diffraction could not be derived because Eq. 5.10 was too difficult to solve analytically even for the unfocused case. We are doubtful that a closed-form solution can be found, if indeed one exists. More resourceful minds may, however, prevail.

Along the same lines, Eq. 5.11 proved difficult to solve in closed form even after substituting $H_1(\rho, z, \omega)$ with $\widehat{H}_1(\rho, z, \omega)$, the Lommel diffraction formulation. Nonetheless, we are somewhat optimistic that a closed-form solution can be found. Despite our less-than-enthusiastic optimism, finding a closed-form solution, if one exists, will take great ingenuity and tenacity.

Finally, we note that the effective backscattering cross-section can be extended to an *effective resolution cell volume*. All that needs to be done is to multiply the autoconvolution diffraction integral in Eq. 5.19 with a weighting function, say $w(z)$, and integrate over z either analytically or numerically. That is,

$$\left|\langle \widehat{H}_2(\omega) \rangle_V\right| \approx \frac{1}{V} \int_z w(z) \left[2\pi \int_0^b \left|\widehat{H}_1(\rho, z, \omega)\right|^2 \rho \, d\rho\right] dz, \qquad (8.1)$$

where the notation should be obvious. This notion of an effective resolution cell volume has been studied elsewhere [75, 83], but Eq. 8.1 might prove to be a useful alternative.

5.   MORE EXPERIMENTS AND ANALYSIS

In Chapter 6, we asked what value of b should be used when implementing autoconvolution diffraction corrections? The notion of an effective backscattering cross-section was introduced to help answer the question, but no definitive answer was postulated. Obtaining a more definitive answer is possible with extensive experimentation and subsequent data analysis. Our suspicion is that the radius of the disk b will have to be varied as a function of depth z.

As we mentioned in Chapter 1, both the mirror-image and autoconvolution interpretations of two-way diffraction have physical merit and mathematical appeal. Experiments designed to compare the two interpretations should be of great interest to the ultrasound community. The experiments should focus not on which interpretation is correct; we have argued that both yield physically meaningful and intuitively appealing results in those instances where they are applicable. Rather, the experiments should be designed to determine the conditions under which the two interpretations hold.

One simple but meaningful two-part experiment immediately pops to mind. The experiment would require the use of two identical transducers. In the first part, the transducers would be set-up for one-way reception. Specifically, the receiver would be placed some distance z in front of the transmitter and centered in its beam. The transmitter would insonify the receiver, and the receiver output would be recorded for later analysis.

In the second part of the experiment, one of the two identical transducers would insonify a large flat plate, and the return from the plate would be recorded. The plate would be placed a distance $z/2$ in front of the transducer and centered it its beam. The radius of the plate should be large enough so that the plate could be considered "infinitely" large, say at least ten times larger than the radius of the transducer. The return from the flat plate would then be rigorously compared with the receiver output obtained from the first part of the experiment. This comparison would give great insight into the nature of one-way and autoconvolution diffraction.

Another suggestion is for an analytical and experimental investigation of linear models of ultrasound. The analytical portion of the investigation would compile a reasonably thorough and authoritative bibliography of linear models of ultrasonic reflection imaging. The bibliography should include (i) a quantitative and qualitative discussion of similarities and differences amongst the models, (ii) rigorous dimensional analysis of terms contained in the models, and (iii) simulations based on a representative sampling of the more cogent models. The experimental portion would compare data obtained from experimental data with results obtained from simulations.

One final recommendation is to study the psychophysical effects of diffraction correction. Trained observers, such as radiologists, sonographers, and researchers in ultrasound, could compare raw and diffraction-corrected images and provide qualitative and quantitative evaluations of the images. The evaluations could be subsequently analyzed using the tools of psychophysics. This is, from a clinical perspective, the most meaningful follow-on work that could be done.

References

[1] R. Altes and W. Faust. A unified method of broad-band echo characterization for diagnostic ultrasound. *IEEE Transactions on Biomedical Engineering*, BME-27(9):520–538, September 1980.

[2] M. Arditi, *et al.* Transient fields of concave annular arrays. *Ultrasonic Imaging*, 3:37–61, 1981.

[3] M. Averkiou, *et al.* A new imaging technique based on the nonlinear properties of tissues. In *1997 IEEE Ultrasonics Symposium*, pages 1561–1565, 1997.

[4] J. Bamber and M. Tristam. Diagnostic ultrasound. In Webb [92], chapter 7. ISBN 0-85274-349-1.

[5] R. Bass. Diffraction effects in the ultrasonic field of a piston source. *Journal of the Acoustical Society of America*, 30(7):602–605, July 1958.

[6] P. Bello. Characterization of randomly time-variant linear channels. In B. Goldberg, editor, *Communications Channels: Characterization and Behavior*, pages 4–37. IEEE Press, New York, NY, 1976. Reprinted from *IEEE Trans. Comm. Sys.*, Vol. CS-11, Dec 1963, pp. 360–393.

[7] G. Benson and O. Kiyohara. Tabulation of some integral functions describing diffraction effects in the ultrasonic field of a circular piston source. *Journal of the Acoustical Society of America*, 55(1):184–185, January 1974.

[8] A. Boivin. On the theory of diffraction by concentric arrays of ring-shaped apertures. *Journal of the Optical Society of America*, 42(1):60–64, January 1952.

[9] G. Box. Robustness in the strategy of scientific model building. In R. Launer and G. Wilkinson, editors, *Robustness in Statistics.* Academic Press, New York, NY, 1979.

[10] O. Bozma and R. Kuc. Characterizing pulses reflected from rough surfaces using ultrasound. *Journal of the Acoustical Society of America*, 89(6):2519–2531, June 1991.

[11] R. Bracewell. *Two-Dimensional Imaging.* Prentice-Hall, Englewood Cliffs, NJ, 1995.

[12] E. Brookner, editor. *Radar Technology.* Lex Book, Lexington, MA, 1996. Orignally published by Artech House.

[13] J. Cardoso and M. Fink. Echographic diffraction filters and the diffraction function for random media through an instantaneous time-frequency approach. *Journal of the Acoustical Society of America*, 90(2, Part 1):1074–1084, August 1991.

[14] D. Cassereau, *et al.* Time deconvolution of diffraction effects— Application to calibration and prediction of transducer waveforms. *Journal of the Acoustical Society of America*, 84(3):1073–1085, September 1988.

[15] J. Chen, *et al.* Tests of backscatter coefficient measurement using broadband pulses. *IEEE Transactions on Ultrasonics, Ferroelectrics, and Frequency Control*, 40(5):603–607, September 1993.

[16] X. Chen, *et al.* Acoustic coupling from a focused transducer to a flat plate and back to the transducer. *Journal of the Acoustical Society of America*, 95(6):3049–3054, June 1994.

[17] Z. Cho, *et al. Foundations of Medical Imaging.* John Wiley & Sons, Inc. New York, NY, 1993.

[18] T. Christopher. Finite amplitude distortion-based inhomogeneous pulse echo ultrasonic imaging. *IEEE Transactions on Ultrasonics, Ferroelectronics, and Frequency Control*, 44(1):125–139, January 1997.

[19] T. Christopher. Experimental investigation of finite amplitude distortion-based,second harmonic pulse echo ultrasonic imaging. *IEEE Transactions on Ultrasonics, Ferroelectronics, and Frequency Control*, 45(1):158–162, January 1998.

[20] I. Claesson and G. Salomonsson. Frequency- and depth-dependent compensation of ultrasonic signals. *IEEE Transactions on Ultrasonics, Ferroelectronics, and Frequency Control*, 35(5):1–6, September 1978.

[21] R. Courant and F. John. *Introduction to Calculus and Analysis*, volume I. Interscience Publishers, New York, NY, 1965.

[22] R. Crochiere. A weighted overlap-add method of short-time fourier analysis/synthesis. *IEEE Transactions on Acoustics, Speech, and Signal Processing*, ASSP-28(1):99–101, February 1980.

[23] C. Daly and N. Rao. The arccos and Lommel diffraction formulations. *Journal of the Acoustical Society of America*, 105(6):3067–3077, June 1999. Copyright 1999, Journal of the Acoustical Society of America, Reprinted with permission.

[24] C. Daly and N. Rao. Spatially averaged impulse response for an unfocused piston transducer: A general derivation and further insights. *Journal of the Acoustical Society of America*, 105(3):1563–1566, March 1999. Copyright 1999, Journal of the Acoustical Society of America. Reprinted with permission.

[25] C. Daly and N. Rao. Time- and frequency-domain descriptions of spatially averaged one-way diffraction for an unfocused piston transducer. *Ultrasonics*, 37(3):209–221, March 1999. Copyright 1999. Reprinted with permission from Elsevier Science.

[26] P. Daponte, *et al.* Detection of echoes using time-frequency analysis techniques. *IEEE Transactions on Instrumentation and Measurements*, 45(1):30–40, February 1996.

[27] J. D'Hooge, *et al.* The calculation of the transient near and far field of a baffled piston using low sampling frequencies. *Journal of the Acoustical Society of America*, 102(1):78–86, July 1997.

[28] R. Dickinson. Reflection and scattering. In C. Hill, editor, *Physical Principles of Medical Ultrasound*, chapter 6. Halstead Press, New York, NY, 1986.

[29] F. Dunn, editor. *Ultrasonic Tissue Characterization*. Springer, Hong Kong, 1996.

[30] M. Fink. Theoretical study of pulsed echographic focusing procedures. In P. Alais and A. Metherell, editors, *Acoustical Imaging*, volume 10. Plenum Press, New York, NY, 1982.

[31] M. Fink and J. Cardoso. Diffraction effects in pulse-echo measurement. *IEEE Transactions on Sonics and Ultrasonics*, SU-31(4):313–329, July 1984.

[32] M. Fink, *et al.* Ultrasonic signal processing for *in vivo* attenuation measurement: Short-time Fourier analysis. *Ultrasonic Imaging*, 5:117–135, 1983.

[33] R. Gagliardi. *Introduction to Communications Engineering.* John Wiley & Sons., New York, NY, second edition, 1988.

[34] J. Gaskill. *Linear Systems, Fourier Transforms, and Optics.* Wiley, New York, NY, 1978.

[35] H. Goldstein. *Classical Mechanics.* Addison-Wesley, Reading, MA, 1950.

[36] J. Goodman. *Introduction to Fourier Optics.* McGraw-Hill Book Co., San Francisco, CA, 1968.

[37] M. Goodsitt, *et al.* Field patterns of pulsed, focused, ultrasonic radiators in attenuating and nonattenuating media. *Journal of the Acoustical Society of America*, 71(2):318–329, February 1982.

[38] A. Gray and G. Mathews. *A Treatise on Bessel Functions and Their Applications to Physics.* Dover Publications, Inc., New York, NY, 1966. First published in 1922.

[39] M. Hamilton and D. Blackstock, editors. *Non-Linear Acoustics: Theory and Applications.* Academic Press, New York, NY, 1998.

[40] G. Harris. Review of transient field theory for a baffled planar piston. *Journal of the Acoustical Society of America*, 70(1):10–19, July 1981.

[41] G. Harris. Transient field of a baffled planar piston transducer having an arbitrary vibration amplitude distribution. *Journal of the Acoustical Society of America*, 70(1):186–204, July 1981.

[42] F. Hlawatsch and G. Bourdreaux-Bartels. Linear and quadratic time-frequency signal representations. *IEEE Signal Processing Magazine*, 9(2):21–67, 1992.

[43] H. Hopkins. The disturbance near the focus of waves of radially non-uniform amplitude. *Proc. Phys. Soc. (London)*, 62B(22):22–32, 1949.

[44] J. Hunt, *et al.* Ultrasound transducers for pulse-echo medical imaging. *IEEE Transactions on Biomedical Engineering*, BME-30(8):453–481, 1983.

[45] H. Huntington, *et al.* Ultrasonic delay lines. I. *Journal of the Franklin Institute*, 245(1), January 1948. See pp. 16–23.

[46] H. Huntley. *The Divine Proportion: A Study in Mathematical Beauty*. Dover Publications, Inc., New York, NY, 1970.

[47] J. Jensen. A model for the propagation and scattering of ultrasound in tissue. *Journal of the Acoustical Society of America*, 89:182–190, January 1991.

[48] J. Jones and S. Leeman. Ultrasonic tissue charaterization and quantitative ultrasound scatter imaging: Methods and approaches. In IEEE Press, editor, *International Workshop on Physics and Engineering in Medical Imaging*, pages 247–258, 1982.

[49] A. Khimunin. Numerical calculation of the diffraction corrections for the precise measurement of ultrasound absorption. *Acustica*, 27:173–181, January 1972.

[50] G. Kino. *Acoustic Waves: Devices, Imaging, & Analog Signal Processing*. Prentice-Hall, Inc., Englewood Cliffs, NJ, 1987.

[51] L. Kinsler, *et al. Fundamentals of Acoustics*. John Wiley & Sons., New York, NY, third edition, 1982.

[52] G. Kossoff. Design of narrow-beamwidth transducers. *Journal of the Acoustical Society of America*, 35(6):905–912, June 1963.

[53] G. Kossoff. Analysis of focusing action of spherically curved transducers. *Ultrasound in Medicine & Biology*, 5:359–365, 1979.

[54] R. Kuc. Modeling acoustic attenuation of soft tissue with a minimum-phase filter. *Ultrasonic Imaging*, 6:24–36, 1984.

[55] R. Kuc and D. Regula. Diffraction effects in reflected ultrasound spectral estimates. *IEEE Transactions on Biomedical Engineering*, BME-31(8):537–545, August 1984.

[56] L. Levi. *Applied Optics: A Guide to Modern Optical System Design*, volume 1. John Wiley & Sons, Inc. New York, NY, 1968.

[57] J. Lim, editor. *Advanced Topics in Signal Processing*. Prentice-Hall, Inc., Englewood Cliffs, NJ, 1988.

[58] M. Linzer, editor. *Ultrasonic Tissue Characterization*, volume 453. National Bureau of Standards, October 1976. A NBS Special Publication.

[59] F. Lizzi, *et al.* Theoretical framework for spectrum analysis in ultrasonic tissue characterization. *Proceedings of the IEEE*, 80(4):520–538, April 1992.

[60] A. Macovski. *Medical Imaging Systems*. Prentice-Hall, Englewood Cliffs, NJ, 1983.

[61] E. Madsen, *et al.* Continuous waves generated by focused radiators. *Journal of the Acoustical Society of America*, 70(5):1508–1517, November 1981.

[62] E. Madsen, *et al.* Method of data reduction for accurate determination of acoustic backscatter coefficient. *Journal of the Acoustical Society of America*, 76(3):913–923, September 1984.

[63] M. Malik, *et al.* Joint-time frequency processing of ultrasonic signals. In D. Thompson and D. Chimenti, editors, *Review of Progress in Quantitative Nondestructive Evaluation*, volume 15, pages 2089–2096. Plenum Press, New York, NY, 1996.

[64] P. Morse and K. Ingard. *Theoretical Acoustics*. McGraw-Hill, New York, NY, 1968.

[65] S. Nawab and T. Quatieri. Short-time fourier transform. In Lim [57], chapter 6.

[66] F. Oberhettinger. On transient solutions of the "baffled piston" problem. *Journal of Research of the National Bureau of Standards B: Mathematics and Mathematical Physics*, 65B(1):1–5, 1961.

[67] H. O'Neil. Theory of focusing radiators. *Journal of the Acoustical Society of America*, 21(5):516–526, September 1949.

[68] A. Oppenheim and R. Shafer. *Discrete-Time Signal Processing*. Prentice Hall, Englewood Cliffs, NJ, 1989.

[69] Panametrics, Inc., Waltham, MA. *Ultrasonic Products for Nondestructive Testing*, 1988.

[70] E. Papadakis and K. Fowler. Broad-band transducers: Radiation field and selected applications. *Journal of the Acoustical Society of America*, 50(3):729–745, 1971.

[71] A. Papoulis. *Systems and Transforms with Applications in Optics.* McGraw-Hill Book Co., New York, NY, 1968. See, in particular, Chapters 5 and 9.

[72] A. Penttinen and M. Luukkala. The impulse response and pressure nearfield of a curved ultrasonic radiator. *Journal of Physics D: Applied Physics*, 9:1547–1557, 1976.

[73] J. Quistgaard. Signal acquisition and processing in medical diagnostic ultrasound. *IEEE Signal Processing Magazine*, pages 21–67, 1997.

[74] S. Rabiner and D. Schaefer. *Digital Processing of Speech Signals.* Prentice-Hall PTR, Upper Saddle River, NJ, 1996.

[75] N. Rao and H. Zhu. Simulation study of changes in ultrasound speckle statistics with the system point spread function. *Journal of the Acoustical Society of America*, 95(2):1161–1164, February 1994.

[76] T. Rhyne. Radiation coupling of a disk to a plane and back or a disk to disk: An exact solution. *Journal of the Acoustical Society of America*, 61(2):318–324, February 1977.

[77] M. Riley. *Speech Time-Frequency Representations.* Kluwer Academic Publishers, Boston, MA, 1989.

[78] D. Robinson, *et al.* Near field transient radiation patterns for circular pistons. *IEEE Transactions on Acoustics, Speech, and Signal Processing*, ASSP-22(6):395–403, December 1974.

[79] P. Rogers and A. Van Buren. An exact expression for the Lommel diffraction correction integral. *Journal of the Acoustical Society of America*, 55(4):728–728, April 1974.

[80] G. Salomonsson and L. Björkman. On the separation of attenuation and texture of tissue using a parametric time-varying network. *IEEE Transactions on Ultrasonics, Ferroelectronics, and Frequency Control*, UFFC-33(3):280–286, May 1986.

[81] H. Seki, *et al.* Diffraction effects in the ultrasonic field of a piston source and their importance in the accurate measurement of attenuation. *Journal of the Acoustical Society of America*, 28(2):230–238, March 1956. There are numerous typos in this article.

[82] K. Shung, *et al. Principles of Medical Imaging.* Academic Press, San Diego, CA, 1992.

[83] R. Sigelmann and J. Reid. Analysis and measurement of ultrasound backscattering from an ensemble of scatterers excited by sine-wave bursts. *Journal of the Acoustical Society of America*, 53(5):1351–1355, 1973.

[84] R. Silverman, *et al.* Measurement of ocular tumor volumes from serial, cross-sectional ultrasound scans. *Retina*, 13(1):69–74, January-March 1993.

[85] M. Skolnik. *Introduction to Radar Systems.* McGraw-Hill Publishing Company, New York, NY, 1980.

[86] P. Stepanishen. Transient radiation from pistons in an infinite planar baffle. *Journal of the Acoustical Society of America*, 49:1627–1638, July 1971.

[87] P. Stepanishen. Pulsed transmit/receive response of ultrasonic piezoelectric transducers. *Journal of the Acoustical Society of America*, 69:1815–1826, June 1981.

[88] J. Thijssen. Echographic image processing. In P. Hawkes, editor, *Advances in Electronics and Electron Physics*, volume 84, pages 317–349. Academic Press, Inc., San Diego, CA, 1992.

[89] Y. Wang and P. Fish. Comparison of Doppler signal analysis techniques for velocity waveform, turbulence and vortex measurement: A simulation study. *Ultrasound in Medicine & Biology*, 22(5):635–649, 1996.

[90] G. Watson. *A Treatise on The Theory of Bessel Functions.* Cambridge University Press, New York, NY, 1980. First published in 1922.

[91] K. Wear, *et al.* Differentiation between acutely ischemic myocardium and zones of completed infarction in dogs on the basis of frequency-dependent backscatter. *Journal of the Acoustical Society of America*, 85(6):2634–2641, June 1989.

[92] S. Webb, editor. *The Physics of Medical Imaging.* IOP Publishing Ltd, Bristol, England, 1988. ISBN 0-85274-349-1.

[93] J. Weight and A. Hayman. Observations of the propagation of very short ultrasonic pulses and their reflection by small targets. *Journal of the Acoustical Society of America*, 63(2):396–404, February 1978.

[94] A. Wheelon. *Tables of Summable Series and Integrals Involving Bessel Functions.* Holden-Day, San Francisco, CA, 1968.

[95] A. Williams, Jr. The piston source at high frequencies. *Journal of the Acoustical Society of America*, 23(1):1–6, January 1951.

[96] A. Williams, Jr. Integrated signal on circular piston receiver centered in a piston beam. *Journal of the Acoustical Society of America*, 48(1):285–289, 1970.

[97] E. Wolf. Light distribution near focus in an error-free diffraction image. *Proceedings of the Royal Society, A*, 204:533–548, 1951.

[98] E. Wolf. The X_n and Y_n functions of Hopkins, occurring in the theory of diffraction. *Journal of the Optical Society of America*, 43(3):218, March 1953.

[99] J. zemanek. Beam behavior within the nearfield of a vibrating piston. *Journal of the Acoustical Society of America*, 49(1, Part 2):181–1084, 1971.

[100] R. Ziemer and W. Tranter. *Principles of Communications: Systems, Modulation, and Noise.* Houghton Mifflin Company, Boston, MA, 1990.

[101] R. Ziemer, *et al. Signals and Systems.* Macmillan Publishing Co., Inc., New York, NY, 1983.

About the Authors

Charles "Chaz" J. Daly was born in Philadelphia, PA in 1957 and is currently a Major in the United States Air Force. He has been serving in the Air Force for over two decades. In his enlisted days, he was a Chinese and Korean linguist at Osan Air Base, Republic of Korea. In 1988, he earned an undergraduate degree, Summa Cum Laude, in Electrical Engineering from Texas A&M University, College Station, TX. He was then commissioned and went to Vandenberg AFB, CA where he worked as a Technical Engineer on Minuteman and Peacekeeper missile systems. Major Daly next earned a Master's Degree in Electrical Engineering from the Air Force Institute of Technology, Dayton, OH. His Master's thesis is entitled *An Analytical and Experimental Investigation of FM-by-Noise Jamming.* He next went to the Air Force Information Warfare Center where he worked on space-support to operational forces and modeling and simulation of satellite communications. Major Daly earned his Ph.D. in Imaging Science from the Rochester Institute of Technology, Rochester, NY in 1998. His dissertation is entitled *The Arccos and Lommel Diffraction Formulations.* He remains on active duty in the USAF.

Navalgund A. H. K. Rao is an associate professor in the Chester F. Carlson Center for Imaging Science at Rochester Institute of Technology, Rochester, NY. His current research interests include linear and non-linear ultrasonic beam propagation modeling for medical applications, signal processing for tissue characterization, and applications of pulse compression techniques. Previously, he has worked as Exploration Geophysicist for the Shell Oil Company and as a National Institute of Health Research Fellow at the University of Colorado Health Science Center in Denver, CO. He is a member of IEEE and AIUM (American Institute of Ultrasound in Medicine).

Index